Student Solutions Manual

for

Devore and Peck's

STATISTICS
The Exploration and
Analysis of Data

Second Edition

John Groves
California Polytechnic State University, San Luis Obispo

Duxbury Press
An Imprint of Wadsworth Publishing Company
A Division of Wadsworth, Inc.

Duxbury Press
An Imprint of Wadsworth Publishing Company
A Division of Wadsworth, Inc.

Printer: Malloy Lithographing, Inc.

Printed in the United States of America

1 2 3 4 5 6 7 8 9 10—97 96 95 94 93

0-534-19616-0

TABLE OF CONTENTS

CHAPTER 1
INTRODUCTION

1.1   Descriptive statistics is made up of those methods whose purpose is to organize and summarize a data set. Inferential statistics refers to those procedures or techniques whose purpose is to generalize or make an inference about the population based on the information in the sample.

1.3   The population of interest is the entire student body (the 15,000 students). The sample consists of the 200 students interviewed.

1.5   The population consists of all single family homes in Pasadena. The sample consists of the 100 homes selected for inspection.

1.7   The population consists of all 5000 bricks in the lot. The sample consists of the 100 bricks selected for inspection.

CHAPTER 2

TABULAR AND PICTORIAL METHODS FOR DESCRIBING DATA

Section 2.1

2.1   a)   numerical (discrete)
        b)   categorical
        c)   numerical (continuous)
        d)   numerical (continuous)
        e)   categorical

2.3   a)   discrete
        b)   continuous
        c)   discrete
        d)   discrete
        e)   continuous
        f)   continuous
        g)   continuous
        h)   discrete

2.5

```
4 | 48,01,50,19,58,17,05,40,02,07,33,92,00,84,24,17,68
5 | 12,01,69,09,99,59,54,53,65,24
6 | 44,47,24,03,34
7 | 30,92,55,91,77,23                    stem: hundreds
8 | 85,56                                leaf: ones
```

2.7

```
64 | 64,70,35,33                         stem: hundreds
65 | 26,83,06,27                         leaf: ones
66 | 14,05,94
67 | 70,70,90,00,98,45,13
68 | 50,73,70,90
69 | 36,27,00,04
70 | 05,40,22,11,51,50
71 | 31,69,68,05,65,13
72 | 09,80
```

The shortest and longest courses are 6433 and 7280 respectively.
If the thousands digits were used as the stem, there would be only
two stems, 6 and 7. If they in turn were broken into 6l, 6h, 7l,
7h, there would be only four groups. Neither situation would yield
a sufficient number of categories to reveal much information. If
the first three leading digits were used as the stems then there
would be eighty six stems, 643 to 728, and this would be too many.

2.9

Marina Del Rey   **Los Angeles-Long Beach**

```
                          41 | 35
                    50    4h | 50,50,60,70,75,95
                    30    5l | 00,00,20,40
                 60,95    5h | 50
         00,27,37,45,49   6l | 00,00,00
      60,60,64,82,82,96   6h | 50
         04,05,05,16      7l |
                          7h |
                          8l |                 stem: dollars
                          8h | 75              leaf: cents
```

The rental rates for Marina del Rey are higher than for the Los
Angeles-Long Beach Harbor. Most of the marinas in Marina del Rey
charge $6.00 or more per foot, while those in the Los Angeles-Long
Beach Harbor charge $6.00 or less per foot. The $8.75 value in the
Los Angeles-Long Beach data set is unusually large. It would be
considered an outlier.

2.11

	Athletes		**Nonathletes**

```
                             │ 0L │ 3,3,4
                       8,8,9 │ 0H │ 6,7,8,9
         0,0,0,2,2,3,3,4,4   │ 1L │ 2,3,3,4,4,4,4,4
5,5,5,5,5,6,6,6,7,7,7,7,8,8,9│ 1H │ 5,5,5,6,7,7,7,8,8,8,8
           0,1,1,1,1,3,4     │ 2L │ 0,2,3,4,4
             6,7,7,8,9,9     │ 2H │ 5,5,5,8,9,9
                             │ 3L │ 0,4,4,4
               5,5,6,7,7,9   │ 3H │ 8
                         0   │ 4L │ 4
                             │ 4H │ 5,8,8
                             │ 5L │ 2
```

Based on the stem and leaf displays, there does not seem to be any
evidence that the proportion of students disqualified is smaller
for nonathletes than for athletes. There appears to be more
variation in the nonathletes, but other than that, both displays
have generally the same shape and location.

2.13 a)

Number of Times	Frequency	Relative Frequency
0	5	.227
1	4	.182
2	3	.136
3	3	.136
4	1	.045
5	1	.045
6	0	.000
7	0	.000
8	2	.091
9	3	.136
n = 22		.998 ≈ 1.0

b) The proportion of shoppers in this study who never bought the brand under investigation is .227, which converts to 22.7%.

c) Half of 9 is 4.5, so to purchase this brand more than half the time means to have purchased 5 or more. The proportion who purchased 5 or more is .045 + .000 + .000 + .091 + .136 = .272, which converts to 27.2%.
The proportion who purchased the brand all the time is .136, which converts to 13.6%.

2.15

Category	Frequency	Relative Frequency	Cum. Rel. Freq.
1	1	.0059	.0059
2	2	.0118	.0177
3	13	.0765	.0942
4	19	.1118	.2060
5	35	.2059	.4119
6	38	.2235	.6354
7	33	.1941	.8295
8	18	.1059	.9354
9	8	.0471	.9825
10	2	.0118	.9943
11	1	.0059	1.0002 ≈ 1.0
n = 170		1.0002 ≈ 1.0	

a) The proportion of observations that are at most 8 is .9354, which converts to 93.54%. The proportion of observations that are at least 8 is 1 - (proportion that are at most seven) which is 1 - .8295 = .1705, which converts to 17.05%.

b) The proportion of letters that contain between 5 and 10 (inclusive) offspring equals (the proportion that have at most 10) minus (the proportion that have at most 4), which is 0.9943 - .2060 = 0.7883, which converts to 78.83%.

Section 2.3

2.17 a) The suggested class intervals would be inappropriate because there are observations whose values are 30.0 and 40.0. Therefore, there is uncertainty in which intervals these observations should be placed. Does 30.0 get placed into 20-30 or 30-40? Does 40.0 get placed into 30-40 or 40-50?

b)
Concentration	Frequency
20 -< 30	1
30 -< 40	8
40 -< 50	8
50 -< 60	6
60 -< 70	16
70 -< 80	7
80 -< 90	2
90 -< 100	2
	n = 50

c) The proportion of the concentration observations that were less than 50 is $\frac{(8+8+1)}{50} = \frac{17}{50} = .34$

The proportion of the concentration observations that were at least 60 is $\frac{(16+7+2+2)}{50} = \frac{27}{50} = .54$

d) No, because of the 16 observations in the class 60 -< 70, there is no information in the frequency distribution to indicate how many are less than 65. To estimate the proportion one could assume that 8 of the 16 were between 60 -< 65. Hence, half of the observations in the 60 -< 70 class are in the lower half, 60 -< 65. Thus, an estimate of the proportion of observations in the sample which are less than 65 would be $\frac{(1+8+8+6+8)}{50} = \frac{31}{50} = .62..$ The actual proportion (from the original data) is $\frac{31}{50} = .62$.

2.19 a)
Class Intervals	Frequency	Relative Frequency
0 -< 100	21	.21
100 -< 200	32	.32
200 -< 300	26	.26
300 -< 400	12	.12
400 -< 500	4	.04
500 -< 600	3	.03
600 -< 700	1	.01
700 -< 800	0	.00
800 -< 900	1	.01
	n = 100	1.00

6

b) Since the second interval of 100 -< 200 is 100 units wide, it is reasonable to approximate the proportion of observations between 110 and 200 to be

$$\frac{(200-110)}{(200-100)}(.32) = \frac{90}{100}(.32) = .288.$$

Hence, the proportion of yarn samples that require a breaking strength of at least 110 cycles is approximately .288 + .26 + .12 + .04 + .03 + .01 + .00 + .01 = .758, which converts to 75.8%. Therefore, 100 - 75.8 = 24.2% would be considered unsatisfactory.

2.21 a) and b)

Class Intervals	Freq.	Rel. Freq.	Cum. Rel. Freq.
0 -< 6	2	.0225	.0225
6 -< 12	10	.1124	.1349
12 -< 18	21	.2360	.3709
18 -< 24	28	.3146	.6855
24 -< 30	22	.2472	.9327
30 -< 36	6	.0674	1.0001 ≈ 1.0
	n = 89	1.0001 ≈ 1.0	

c) (Rel. Freq. for 12 -< 18)
= (Cum. Rel. Freq. for < 18) -
  (Cum. Rel. Freq. for < 12)
= .3709 - .1349
= .2360.

d) The proportion that had pacemakers that did not malfunction within the first year equals 1 minus the proportion that had pacemakers that malfunctioned within the first year (12 months), which is 1 - .1349 = .8651, which converts to 86.51%.

e) The proportion that required replacement between one and two years after implantation is equal to the proportion that had to be replaced within the first 2 years (24 months) minus the proportion that had to be replaced within the first year (12 months). This is 0.6855 - 0.1349 = 0.5506, which converts to 55.06%.

f) The proportion that lasted less than 18 months is .3709, which converts to 37.09%, and the proportion that lasted less than 24 months is .6855, which converts to 68.55%. Thus, the time at which about 50% of the pacemakers failed is somewhere between 18 and 24 months. A more precise estimate can be found as follows.

$$\frac{(.50 - .3709)}{.3146} = \frac{x}{6} \rightarrow x = \frac{6(.50 - .3709)}{.3146} = 2.46$$

So the time at which about 50% of the pacemakers had failed is 18 + 2.46 or 20.46 months.

g) $\dfrac{(.9327 - .9)}{.2472} = \dfrac{x}{6} \rightarrow x = \dfrac{6(.9327 - .9)}{.2472} = .79$

So an estimate of the time at which only 10% of the pacemakers initially implanted were still functioning is 30 - .79 = 29.21 months.

2.23

Fuel Efficiency	Cum. Rel. Freq.
27.0 -< 27.5	.09
27.5 -< 28.0	.23
28.0 -< 28.5	.38
28.5 -< 29.0	.35 ⇐
29.0 -< 29.5	.72
29.5 -< 30.0	.80
30.0 -< 30.5	.93
30.5 -< 31.0	1.00

The cumulative relative frequencies must be an increasing sequence of numbers. Therefore, to report .09, .23, .38, then .35, indicates that an error in reporting was made.

2.25

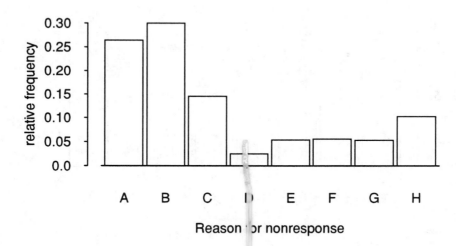

Reason for nonresponse

2.27

Sport	Frequency	Rel. Freq.
Touch Football (TF)	38	0.226
Soccer (SO)	24	0.143
Basketball (BK)	19	0.113
Baseball/Softball (BA)	11	0.065
Jogging/Running (JR)	11	0.065
Bicycling (BI)	11	0.065
Volleyball (VO)	7	0.042
Others (OT)	47	0.028
	n = 168	0.999

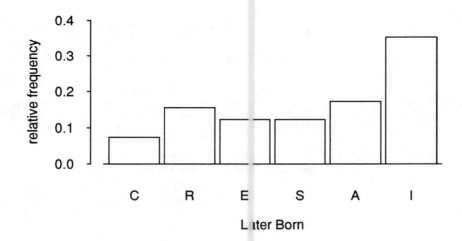

Later Born

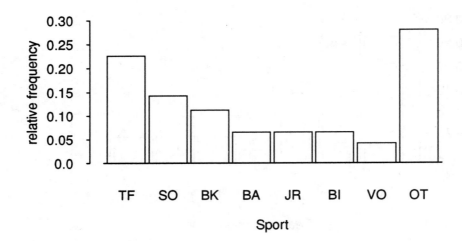

2.29   a)

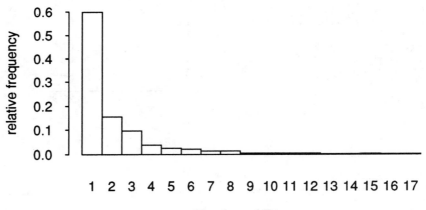

b)   No, since there would not be any way of determining the largest observation, the right side boundary point of the rectangle for the class "≥ 15" could not be determined.

c)   Yes, because now you know the largest observation is 17 and the right side boundary point would be placed at 17.5.

2.31 a)

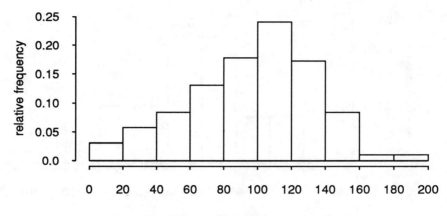

Distance Traveled (1000's miles)

b)   The estimate of the proportion of all buses of this type
      that operate for more than 100,000 miles before the first
      major motor failure would be

      .241 + .173 + .084 + .01 + .01 = .518,

      which converts to 51.8%.

c)   The estimate of the proportion of all buses of this type
      which have their first major motor failure after operating
      for between 50,000 and 125,000 miles is
      .5(.084) + .131 + .178 + .241 + .25(.173)=.635.

2.33

Class Interval	Frequency	Rel. Freq.
30 -< 50	2	.04
50 -< 70	4	.08
70 -< 90	4	.08
90 -< 110	11	.22
110 -< 130	12	.24
130 -< 150	3	.06
150 -< 170	5	.10
170 -< 190	4	.08
190 -< 210	3	.06
210 -< 230	1	.02
230 -< 250	1	.02
	n = 50	1.00

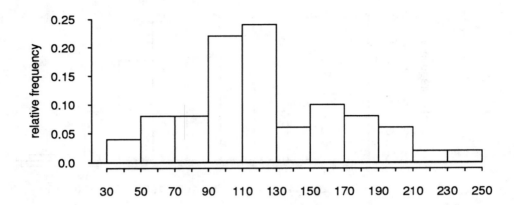

Average Payment per Person

2.35 a) If the exam is quite easy, then there would be a large number of high scores with a small number of low scores. The resulting histogram would be negatively skewed.

b) If the exam is quite hard, then there would be a large number of low scores with a small number of high scores. The resulting histogram would be positively skewed.

c) The students with the better math skills would score high, while those with poor math skills would score low. This would result in basically two groups and thus the resulting histogram would be bimodal.

2.37 **Class Intervals**

Class Intervals	I	II	III	IV
100 -< 120	5	5	35	20
120 -< 140	10	7	15	10
140 -< 160	40	10	10	4
160 -< 180	10	15	5	25
180 -< 200	5	33	5	11

a) The histogram associated with frequency set I would be symmetric.

b) The histogram associated with frequency set II would be negatively skewed.

c) The histogram associated with frequency set III would be positively skewed.

d) The histogram associated with frequency set IV would be bimodal.

2.39 a)

Class Interval	Frequency	Relative Frequency
54 -< 56	1	0.025
56 -< 58	0	0.000
58 -< 60	6	0.150
60 -< 62	8	0.200
62 -< 64	12	0.300
64 -< 66	8	0.200
66 -< 68	3	0.075
68 -< 70	2	0.050
n = 40		1.000

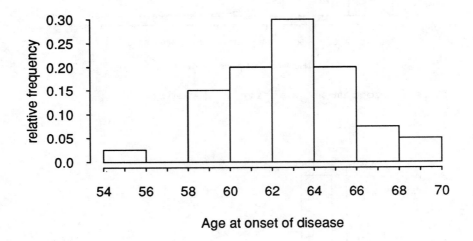

b)   This histogram is not skewed very much, however, there appears to be some negative skewness to it. It is a tough call to make based solely on the graph.

c)   Based on the answer of part (b), there appears to be little skewness to the histogram. Therefore, transforming this data would most likely not be productive.

2.41 a)

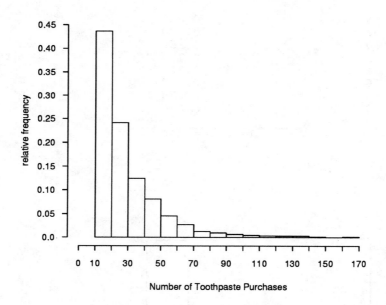

This histogram is positively skewed.

b)

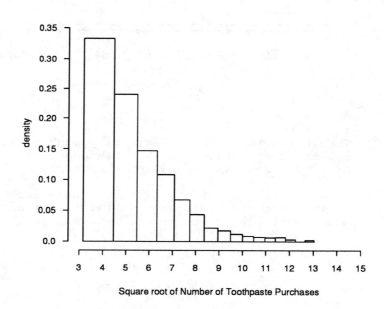

This histogram is still positively skewed.

2.43

Category	Frequency	Rel. Freq.
Clerical (C)	4	.08
Manager (M)	6	.12
Professional (P)	11	.22
Retired (R)	18	.36
Sales(S)	8	.16
Tradesman (T)	2	.04
Other (O)	1	.02
	n = 50	1.00

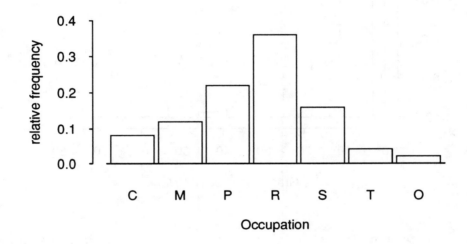

2.45 a)

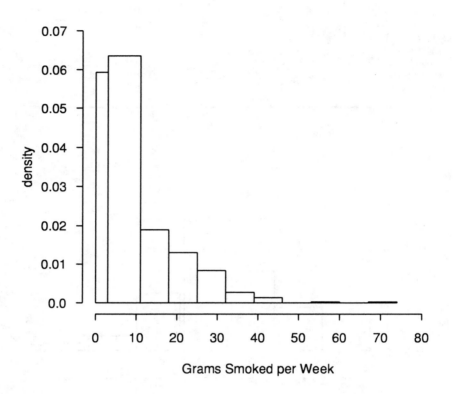

b)   The proportion of respondents who smoked 25 or more grams per week is $\frac{48}{529}$, or using the relative frequencies .0586 + .0189 + .0095 + .0019 + .0019 = .0908, which converts to 9.08%.

c)   The approximate proportion who smoked between 15 and 18 grams per week is

$$\frac{(18-15)}{(18-11)}(.1323) = \frac{3}{7}(.1323) = .0567$$

Thus, the approximate proportion who smoked more than 15 grams per week is .0567 + .0907 + .0586 + .0189 + .0095 + .0019 + .0019 = .2382, which converts to 23.82%.

Supplementary Exercises

2.47 a)

```
 0 | 5 5 8 3 7 8 6 7 8 6 7 2 5
 1 | 1 0 0 5 5 0 5 3 0 5 5
 2 | 0 2 7 5 0
 3 | 3 0 3 0
 4 | 0 0 7 5
 5 | 4 0 4
 6 |
 7 | 0 5
 8 | 8
 9 | 0                              stem:  ones
10 | 3                              leaf:  tenths
```

HI: 22.0 and 24.5

b)

Classes	Frequency	Rel. Freq.	Height
0 -< 2	24	.5106	.2553
2 -< 4	9	.1915	.0957
4 -< 6	7	.1489	.0745
6 -< 10	4	.0851	.0213
10 -< 20	1	.0213	.0021
20 -< 30	2	.0426	.0043
	n = 47	1.0000	

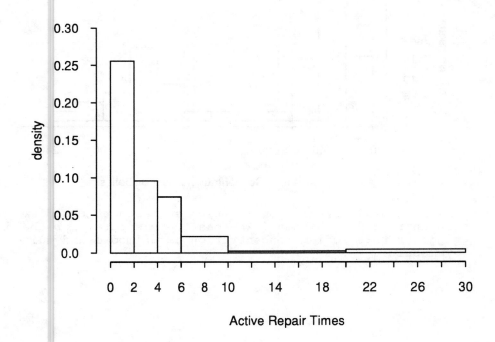

17

2.49

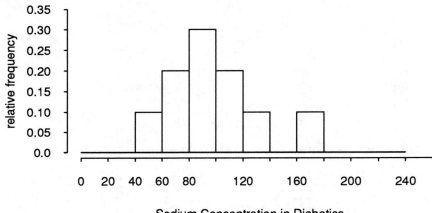

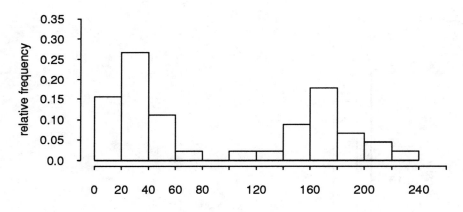

The histogram for the nondiabetics suggests that sodium concentration is bimodal, whereas the histogram for the diabetics suggests that sodium concentration is unimodal and quite symmetric.

2.51

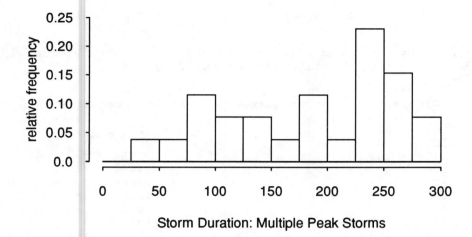

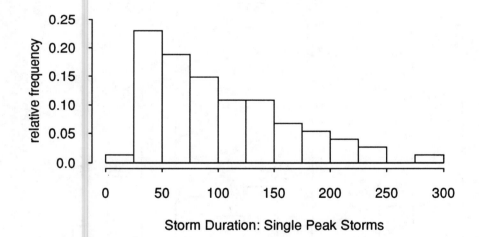

The only apparent similarity is that both distributions range from 0 to 300 minutes. The histogram for single-peak storms is skewed positively with a rather even decline as the minutes of duration increase. The durations of most single-peak storms were between 25 and 100 minutes. The histogram for multiple-peak storms is skewed negatively with a lot of variation in height of the rectangles. Most multiple-peak storms lasted either between 75 to 100 minutes or 225 to 300 minutes. This suggests that the distribution may be bimodal.

2.53　A stem and leaf display for the data results in

```
3L | 1 2 3 3 3 4 4 4
3H | 5 7 7 9 9
4L | 0 0 0 0 0 1 2 3 3 4 4
4H | 5 5 5 6 6 6 6 6 6 6 7 7 7 7 7 7 8 8 8 8 8 8 9
5L | 0 0
```

where a stem of 3 and a leaf of 1 denote a number between 310 and 319. A frequency distribution for the data is:

Classes	Frequency	Relative Frequency
300 -< 320	1	.0204
320 -< 340	4	.0816
340 -< 360	4	.0816
360 -< 380	2	.0408
380 -< 400	2	.0408
400 -< 420	6	.1224
420 -< 440	3	.0612
440 -< 460	5	.1020
460 -< 480	13	.2653
480 -< 500	7	.1429
500 -< 520	2	.0408
	n = 49	.9998

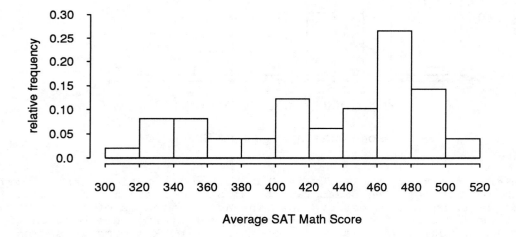

2.55

Class	Freq.	Rel. Freq.
.175 -< .225	4	.0727
.225 -< .275	2	.0364
.275 -< .325	16	.2909
.325 -< .375	15	.2727
.375 -< .425	9	.1636
.425 -< .475	6	.1091
.475 -< .525	2	.0364
.525 -< .575	0	.0000
.575 -< .625	1	.0182
	n = 55	1.0000

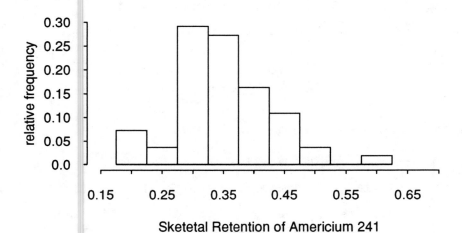

Sketetal Retention of Americium 241

The histogram is skewed slightly in the positive direction. The majority of the observations are in the center.

2.57   The transformation on Mn has been very successful in obtaining a symmetric distribution. The transformation on Zn appears to have produced a bimodal distribution. It certainly is not as symmetric as the transformed Mn. The transformation on Cu has been reasonably successful but not as much as for Mn.

Section 3.1

3.1   a) $\overline{x} = \dfrac{928}{17} = 54.59$

      b)   The data arranged in ascending order is:

           48,50,50,50,50,53,53,54,55,55,55,56,56,58,59,63,63.

           The median is the middle number and its value is 55. The mean is slightly less in value than the median.

      c) $\overline{x} = \dfrac{1550}{17} = 91.18$

           The average zinc concentration at the 17 Texaco drilling sites is 91.18 parts per million.

3.3   The mean for the sample of 14 values is

$$\frac{[13(119.7692)+159]}{14} = \frac{[1557+159]}{14}$$
$$= \frac{1716}{14} = 122.57$$

3.5   a)   The sample mean equals $\dfrac{222}{16} = 13.875$.

      b)   The ordered sample data is:

           3, 6, 7, 10, 12, 13, 13, 14, 14, 14, 16, 18, 19, 19, 22, 22.

           The median is the average of the eighth and ninth observations. It is therefore equal to $\dfrac{(14+14)}{2} = 14$.

      c)   Deleting the smallest value which is 3, the largest value which is 22, and then averaging, yields the 6.25% trimmed mean of 197/14 = 14.07. This yields a slightly more representative value because most of the data is between 10 and 20, and the effect of the smallest value (3) is eliminated.

      d)   The data arranged in ascending order is:

           5, 8, 8, 9, 9, 9, 10, 12, 12, 13, 13, 13, 14, 16, 16, 17.

           The mean is 184/16 = 11.5 and the median is (12+12)/2 = 12. Both of these values are smaller than their respective counterparts in part (a). The data suggests that the ozone levels are typically smaller at the Coliseum than at East Los Angeles College.

3.7   This statement could be correct if there were a small group of residents with very large wages. This group would make the average wage very large and thus a large percentage (perhaps even as large as 65%) could have wages less than the average wage.

3.9   The ordered values are:

Diet 1: 5 7 12 13 15
Diet 2: 3 5 10 11 13

Since each ordered observation in diet 1 is larger than the corresponding ordered observation in diet 2, it follows that the total for diet 1 is larger than the total for diet 2. Since the sample sizes are the same, it then follows that the average weight gain for diet 1 is larger than that for diet 2.

3.11   The ordered data set is 29, 30, 31, x_5, 49. The sample median is 31. The sample mean is $\bar{x} = \dfrac{[29+30+31+(31+k)+49]}{5} = \dfrac{(170+k)}{5} = 34+\dfrac{k}{5}$

Since $k \geq 0$, the sample mean, $34 + k/5$, will be larger than the sample median, which is 31.

3.13   The median and trimmed mean (trimming percentage of at least 20) can be calculated.

sample median = $(57+79)/2 = 136/2 = 68$
20% trimmed mean = $(35 + 48 + 57 + 79 + 86 + 92)/6 = 397/6 = 66.17$

3.15   a)   average over sections = $\dfrac{20+100}{2} = 120/2 = 60$

       b)   average over students =

$$\dfrac{20(20)+100(100)}{20+100} = \dfrac{400+10000}{120} = 10400/120 = 86.67.$$

       c)   Responses will vary.

3.17 Σx = (14.6 + 14.3 + 15.1 + 12.7 + 11.8 + 13.4 + 13.8)
 = 95.7

 Σx^2 = (14.6² + 14.3² + 15.1² + ... + 13.8²)
 = (213.16 + 204.49 + 228.01 + ... + 190.44)
 = 1316.19

$$s^2 = \frac{1316.19 - \frac{(95.7)^2}{7}}{7-1} = \frac{1316.19 - 1308.35}{6} = \frac{7.84}{6}$$
 = 1.307

$s = \sqrt{1.307}$ = 1.143

3.19 a) Set 1: 2, 3, 7, 11, 12 ; $\overline{x}$ = 7 and s = 4.528
 Set 2: 5, 6, 7, 8, 9 ; $\overline{x}$ = 7 and s = 1.581

 b) Set 1: 2, 3, 4, 5, 6 ; $\overline{x}$ = 4 and s = 1.581
 Set 2: 4, 5, 6, 7, 8 ; $\overline{x}$ = 6 and s = 1.581

3.21 $s^2 = \dfrac{1004.4 - \frac{(100.2)^2}{10}}{10-1} = \dfrac{1004.4 - 1004.004}{9} = \dfrac{.396}{9} = .044$

$s = \sqrt{.044}$ = .2098

3.23 Multiplying each data point by 10 yields

x	$x - \overline{x}$	$(x-\overline{x})^2$
620	136.36364	18595.04132
230	-253.63636	64331.40496
270	-213.63636	45640.49587
560	76.36364	5831.40496
520	36.36364	1322.31405
340	-143.63636	20631.40496
420	-63.63636	4049.58678
400	-83.63636	6995.04132
680	196.36364	38558.67769
450	-33.63636	1131.40496
830	346.36364	119967.76860
	0.00004	327054.54545

$\overline{x}$ = 483.63636
$s^2 = \dfrac{(327054.54545)}{10} = 32705.4545$
$s = \sqrt{32705.4545}$ = 180.846

The standard deviation for the new data set is 10 times larger than the standard deviation for the original data set.

3.25   The ordered sample is:

   lower half: 2.34, 2.43, 2.62, 2.74, 2.74, 2.75, 2.78, 3.01, 3.46

   upper half: 3.46, 3.56, 3.65, 3.85, 3.88, 3.93, 4.21, 4.33, 4.52

   a)   lower quartile = 2.74   upper quartile = 3.88

   b)   iqr = 3.88 - 2.74 = 1.14

   c)   The value of the iqr would not be affected by this change
        since both 5.33 and 5.52 are larger than the upper quartile.

   d)   The observation 2.34 could be increased to as much as 2.74
        (the value of the lower quartile) without affecting the iqr.

   e)   If an 18th observation, 4.60, is added to the data, then the
        lower quartile = 2.74, the upper quartile = 3.93 and the iqr
        = 3.93 - 2.74 = 1.19.

3.27   a)   For sample 1, $\overline{x}$ = 7.81 and s = .39847
            For sample 2, $\overline{x}$ = 49.68 and s = 1.73897

       b)   For sample 1, CV = (100)(.39847)/7.81 = 5.10
            For sample 2, CV = (100)(1.73897)/49.68 = 3.50

3.29 a)   The value 57 is one standard deviation above the mean. The value 27 is one standard deviation below the mean. By the empirical rule, roughly 68% of the vehicle speeds were between 27 and 57.

b)   From part (a) it is determined that 100% - 68% = 32% were either less than 27 or greater than 57. Because the normal curve is symmetric, this allows us to conclude that half of the 32% (which is 16%) falls above 57. Therefore, an estimate of the percentage of fatal automobile accidents that occurred at speeds over 57 mph is 16%.

3.31 Since the histogram is well approximated by a normal curve, the empirical rule will be used to obtain answers for part a) - c).

a)   Because 2500 is 1 standard deviation below the mean and 3500 is 1 standard deviation above the mean, about 68% of the sample observations are between 2500 and 3500.

b)   Since both 2000 and 4000 are 2 standard deviations from the mean, approximately 95% of the observations are between 2000 and 4000. Therefore about 5% are outside the interval from 2000 to 4000.

c)   Since 95% of the observations are between 2000 and 4000 and about 68% are between 2500 and 3500, there is about 95-68 = 27% between 2000 and 2500 or 3500 and 4000. Half of those, 27/2 = 13.5%, would be in the region from 2000 to 2500.

d)   When applied to a normal curve, Chebyshev's rule is quite conservative. That is, the percentages in various regions of the normal curve are quite a bit larger than the values given by Chebyshev's rule.

3.33 a)   lower half: 2.62, 2.83, 2.91, 3.49, 3.58, 3.58, 3.59,
                         3.84, 3.86, 3.90, 4.11, 4.25, 4.27

The lower quartile = 3.59.

upper half: 4.36, 4.41, 4.43, 4.46, 4.53, 4.58, 4.65,
              4.75, 4.78, 5.21, 6.00, 6.49, 6.79

The upper quartile = 4.65 and the iqr = 4.65 - 3.59 = 1.06.

b)   In order to be an outlier, an observation would have to be smaller than 3.59 - 1.5(1.06) = 2.00 or larger than 4.65 + 1.5(1.06) = 6.24. To be an extreme outlier, an observation would have to be smaller than 3.59 - 3(1.06) = 0.41 or larger than 4.65 + 3(1.06) = 7.83. There are two mild outliers, namely 6.49 and 6.79. There are no extreme outliers.

c)

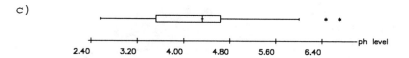

3.35 For the first test the students z-score is $\frac{(625-475)}{100}$ = 1.5 and for the second test it is $\frac{(45-30)}{8}$ = 1.875. Since the students' z-score is larger for the second test than for the first test, the student's performance was better on the second exam.

3.37 a)

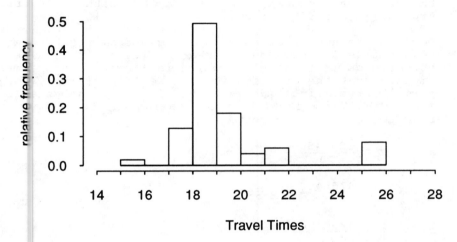

b) i) The 86th percentile is approximately 21, because the cumulative relative frequency for the 20 -< 21 class is .86.

ii) The 15th percentile is approximately 18, because the cumulative relative frequency for the 17 -< 18 class is .15.

iii) The 90th percentile is approximately 21.67, because the cumulative relative frequency up to 21 is 86%, up to 22 is 92%, and 21.67 is two-thirds of the way across the class of 21 -< 22. The value two-thirds was chosen because 90% is two-thirds of the way between 86% and 92%.

iv) The 95th percentile will be in the class 25 -< 26. The cumulative relative frequency at 25 is .92. The 95th percentile will be 3/8 of the way across the 25 -< 26 class because $\frac{(95-92)}{8} = \frac{3}{8}$. Thus, the 95th percentile is 25.375.

v) The 10th percentile will be $\frac{(10-2)}{13} = \frac{8}{13}$ of the way across the 17 -< 18 class. Hence, the value of the 10th percentile is 17.615.

3.39   The recorded weight will be within 1/4 ounces of the true weight
if the recorded weight is between 49.75 and 50.25 ounces. Now,

$$\frac{(50.25-49.5)}{.1} = 7.5 \quad \text{and} \quad \frac{(49.75-49.5)}{.1} = 2.5$$

Also, at least $1 - 1/(2.5)^2 = 84\%$ of the time the recorded weight
will be between 49.25 and 49.75. This means that the recorded
weight will exceed 49.75 no more than 16% of the time. This
implies that the recorded weight will be between 49.75 and 50.25
no more than 16% of the time. That is, the proportion of the time
that the scale showed a weight that was within 1/4 ounce of the
true weight of 50 ounces is no more than .16.

3.41   Because the number of answers changed from right to wrong cannot
be negative and because the mean is 1.4 and the value of the
standard deviation is 1.5, which is larger than the mean, this
implies that the distribution is skewed positively and is not a
normal curve. Since $(6-1.4)/1.5 = 3.07$, by Chebyshev's rule, at
most $1/(3.07)^2 = 10.6\%$ of those taking the test changed at least 6
from correct to incorrect.

3.43   a) $z = \dfrac{320-450}{70} = -1.86$

b) $z = \dfrac{475-450}{70} = 0.36$

c) $z = \dfrac{420-450}{70} = -0.43$

d) $z = \dfrac{610-450}{70} = 2.29$

3.45   The z-score associated with the first stimulus value is (4.2-
6.0)/1.2 = -1.5 and for the second stimulus value it is (1.8-
3.6)/.8 = -2.25. Since the z-score associated with the second
stimulus reading is more negative (-2.25 is less than -1.5), you
are reacting more quickly (relatively) to the second stimulus than
to the first.

3.47   The data arranged in ascending order is:

x	$x - \bar{x}$	$(x-\bar{x})^2$
0.483	-1.185091	1.404441
0.598	-1.070091	1.145095
0.684	-0.984091	0.968435
0.924	-0.744091	0.553671
1.038	-0.630091	0.397015
1.285	-0.383091	0.146759
1.497	-0.171091	0.029272
2.540	0.871909	0.760225
2.650	0.981909	0.964145
3.130	1.461909	2.137178
3.520	1.851909	3.429567
	-0.000001	11.935803

The mean equals $(18.349)/11 = 1.668091$ and the median = 1.285. The variance equals $(11.935803)/10 = 1.1935803$ and the standard deviation equals $\sqrt{1.1935803} = 1.092511$.

3.49   The data arranged in ascending order is:

x	$x - \bar{x}$	$(x-\bar{x})^2$
54	-49.833	2483.4
55	-48.833	2384.7
56	-47.833	2288.0
60	-43.833	1921.4
60	-43.833	1921.4
60	-43.833	1921.4
105	1.167	1.4
120	16.167	261.4
135	31.167	971.4
140	36.167	1308.0
154	50.167	2516.7
247	143.167	20496.7
1246	0.004	38475.9

a)   The sample mean is $1246/12 = 103.83$ and the sample median is $(60 + 105)/2 = 82.5$. The sample variance

$$s^2 = 38475.9/11 = 3497.8091$$

and the sample standard deviation

$$s = 59.14.$$

The average age at death for this group of children is 103.83 days. Half of the children die in less than 82.5 days and the other half after living 82.5 days or longer. The typical difference of age at death from the mean is 59.14 days.

b)   The lower quartile is (56 + 60)/2 = 58 and the upper
     quartile is (135 + 140)/2 = 137.5. The iqr equals 137.5 - 58
     = 79.5
     upper quartile + 1.5(iqr)=137.5+1.5(79.5)=256.75
     Since 247 is less than 256.75, the value 247 is not a mild
     or extreme outlier.

c)

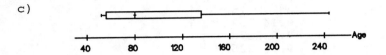

3.51   The data arranged in descending order is:

x	$x - \bar{x}$	$(x-\bar{x})^2$
8.53	0.524000	0.274576
8.52	0.514001	0.264197
8.01	0.004001	0.000016
7.99	-0.016000	0.000256
7.93	-0.076000	0.005776
7.89	-0.116000	0.013456
7.85	-0.156000	0.024336
7.82	-0.185999	0.034596
7.80	-0.205999	0.042436
7.72	-0.286000	0.081796
80.06	0.000004	0.741441

a)   The sample mean is 80.06/10 = 8.006. The sample variance is
     s^2 = .741441/9 = .0824 and the sample standard deviation is
     s = $\sqrt{.0824}$ = .287. The average soil pH level is 8.006. The
     average squared deviation of the pH level is .0824 and the
     typical difference from the mean for pH levels is .287.

b)   The 10% trimmed mean is 63.81/8 = 7.976. The sample median
     is (7.89 + 7.93)/2 = 7.91. The trimmed mean is quite close
     to the mean, but the median is about 1/3 of a standard
     deviation below the mean.

c)   The upper quartile is 8.01, the lower quartile is 7.82, and
     the iqr = 8.01 - 7.82 = 0.19.

d)

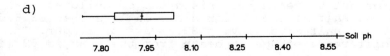

30

3.53

x	$x - \bar{x}$	$(x-\bar{x})^2$
18	-4.15	17.22
18	-4.15	17.22
25	2.85	8.12
19	-3.15	9.92
23	0.85	0.72
20	-2.15	4.62
69	46.85	2194.92
18	-4.15	17.22
21	-1.15	1.32
18	-4.15	17.22
18	-4.15	17.22
20	-2.15	4.62
18	-4.15	17.22
18	-4.15	17.22
20	-2.15	4.62
18	-4.15	17.22
19	-3.15	9.92
28	5.85	34.22
17	-5.15	26.52
18	-4.15	17.22
443	0.00	2454.48

a) $\bar{x} = \dfrac{443}{20} = 22.15$

$s^2 = 2454.48/19 = 129.183$

$s = \sqrt{129.183} = 11.366$

b)   The 10% trimmed mean is calculated by eliminating the two largest values (69 and 28) and the two smallest values (17 and 18). The trimmed mean equals 311/16 = 19.4375. It is a better measure of location for this data set since it eliminates a very large value (69) from the calculation. It is almost 3 units smaller than the average.

c)   The upper quartile is (21+20)/2 = 20.5, the lower quartile is (18+18)/2 = 18, and the iqr = 20.5-18=2.5.

d)   upper quartile + 1.5(iqr) = 20.5+1.5(2.5)=24.25
upper quartile + 3.0(iqr) = 20.5+3(2.5)=28.00
The values 25 and 28 are mild outliers and 69 is an extreme outlier.

e)

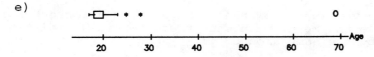

3.55 a)   If one observation is deleted at each end, the trimming percentage would be 1/15 = .0667 or 6.67%. If two observations are deleted at each end, the trimming percentage would be 2/15 = .1333 or 13.33%.

b)   The ordered data is 7.9, 8.5, 8.8, 9.2, 9.3, 9.6, 9.8, 10.5, 10.7, 11.0, 12.1, 12.2, 13.2, 13.7, 16.6. The 6.67% trimmed mean is (8.5+8.8+...+13.7)/13 = 138.6/13 = 10.66. The 13.33% trimmed mean is

$$(8.8+9.2+...+13.2)/11 = 116.4/11 = 10.58$$

c)   To calculate a 10% trimmed mean, one might interpolate linearly between the two computed trimmed means in b). The resulting value would be

$$\frac{10.58-10.66}{.1333-.0667} = \frac{x-10.66}{.10-.0667} \rightarrow \frac{-.08}{.0667} = \frac{x-10.66}{.0333}$$

$$x = 10.66+.0333\left(\frac{-.08}{.0667}\right) = 10.66+.5(-.08) = 10.62$$

3.57 $\Sigma x = 268.8$ $\Sigma x^2 = 2756.54$

$$\overline{x} = \frac{268.8}{27} = 9.956$$

Median value = 10.6

$$s^2 = \frac{2756.54-\dfrac{(268.8)^2}{27}}{26} = \frac{2756.54-2676.053}{26}$$

$$= \frac{80.487}{26} = 3.096$$

$$s = \sqrt{3.096} = 1.759$$

A stem and leaf display is

```
 6 | 3 4
 7 | 1 7
 8 | 4 5 8 9
 9 | 0 1
10 | 0 1 2 6 6 7 7 8 9
11 | 1 2 2 4 9 9
12 | 2
13 | 1
```
                                         stem: ones
                                         leaf: tenths

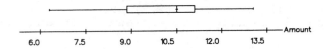

3.59 a)   If for every value below the mean there is a corresponding value above the mean, such that these two values differ from the mean by the same amount.

b)   When the average of the two deleted values equals the value of the trimmed mean.

3.61 $$\overline{x}_w = \frac{120000(268)+30000(210)+45000(220)}{120000+30000+45000} = \$248$$

3.61 $\overline{x}_w = \dfrac{120000(268) + 30000(210) + 45000(220)}{120000 + 30000 + 45000} = \$248$

3.63 a) The 16th and 84th percentiles are the same distance from the mean but on opposite sides. Since 80 is 20 units below 100, the 84th percentile is 20 units above 100, which would be at 120.

b) Since 84-16 = 68, roughly 68% of the scores are between 80 and 120. Thus, by the empirical rule, 120 is one standard deviation above the mean, so the standard deviation has value 20.

c) The value 90 would have a z-score of (90-100)/20 = -1/2.

d) The value 140 would be two standard deviations above the mean and thus roughly .5(5%)=2.5% of the scores would be larger than 140. Hence 140 would be the 97.5 percentile.

e) The value 40 is three standard deviations below the mean, so only about .5(.3%) or .15% of the scores are below 40. There would not be many scores below 40.

4.1

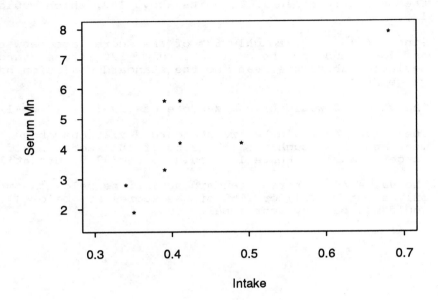

The plot suggests that as intake increases, there is a strong tendency
for Serum Mn to increase also.

4.3   a)   There is not a deterministic relationship between x and y.
           This can be determined by the fact that there are two data
           points, (100, 222) and (100, 241), which have the same x-
           value but different y-values.

b)

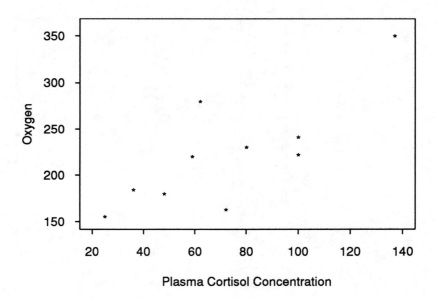

c) There appears to be a tendency for oxygen consumption rate to increase as plasma cortisol concentration increases.

4.5 a)

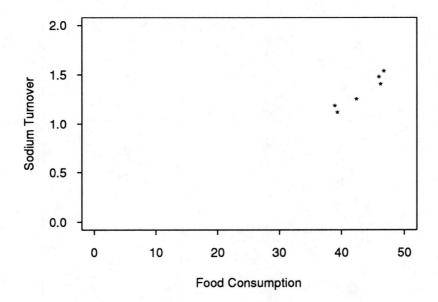

b)

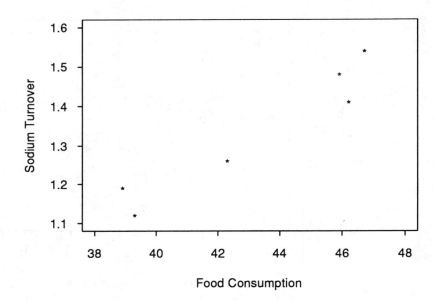

The plot in (b) is more appealing than the plot in (a) because the overall pattern is more apparent.

c)   The pattern in the plot in (b) is generally that of a linear relationship with a positive slope.

4.7   a)   A positive correlation would be expected, since as temperature increases cooling costs would also increase.

       b)   A negative correlation would be expected, since as interest rates climb fewer people would be submitting applications for loans.

       c)   A positive correlation would be expected, since husbands and wives tend to have jobs in similar or related classifications. That is, a spouse would be reluctant to take a low-paying job if the other spouse had a high-paying job.

       d)   No correlation would be expected, because those people with a particular I.Q. level would have heights ranging from short to tall.

       e)   A positive correlation would be expected, since people who are taller tend to have larger feet and people who are shorter tend to have smaller feet.

       f)   A weak to moderate positive correlation would be expected. There are some who do well on both, some who do poorly on both, and some who do well on one but not the other. It is perhaps the case that those who score similarly on both tests outnumber those who don't.

       g)   A negative correlation would be expected, since there is a fixed amount of time and as time spent on homework increases, time in watching television would decrease.

       h)   No correlation overall, because for small or substantially large amounts of fertilizer yield would be small.

4.9

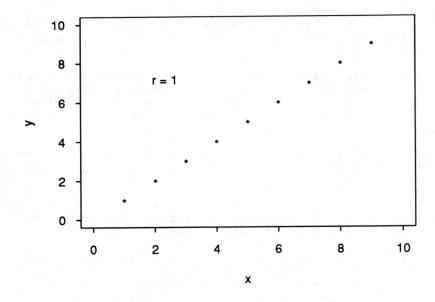

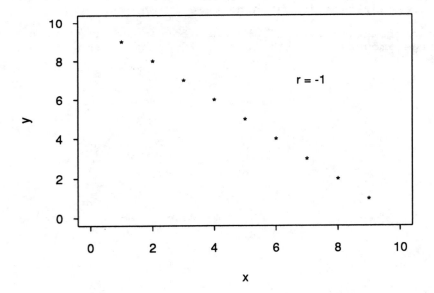

4.11 $n = 16$, $\sum x = 160.5$, $\sum x^2 = 1997.19$, $\sum y = 53.9$, $\sum y^2 = 321.55$, $\sum xy = 623.35$

$$\sum x^2 - \frac{(\sum x)^2}{n} = 1997.19 - \frac{(160.5)^2}{16} = 387.174$$

$$\sum y^2 - \frac{(\sum y)^2}{n} = 321.55 - \frac{(53.9)^2}{16} = 139.974$$

$$\sum xy - \frac{(\sum x)(\sum y)}{n} = 623.35 - \frac{(160.5)(53.9)}{16} = 82.666$$

$$r = \frac{82.666}{\sqrt{(387.174)(139.974)}} = .355$$

Since $r = .355$, the correlation between carbon monoxide concentration and benzo(a)pyrene concentration would be described as a weak positive correlation. Therefore, the relationship cannot be very accurately modeled by a straight line.

4.13   a) $\sum x^2 - \frac{(\sum x)^2}{n} = 251970 - \frac{(1950)^2}{18} = 40720$

$$\sum y^2 - \frac{(\sum y)^2}{n} = 130.6074 - \frac{(47.92)^2}{18} = 3.0337$$

$$\sum xy - \frac{(\sum x)(\sum y)}{n} = 5530.92 - \frac{(1950)(47.92)}{18}$$
$$= 339.5867$$

$$r = \frac{339.5867}{\sqrt{(40720)(3.0337)}} = .966$$

   b)   Since r is quite close to +1, it may be that the relationship between the two variables is much like a linear relationship.

   c)   Since r is positive, as sheer force increases percent fiber dry weight increases.

4.15   No, because x, artist, is not a numerical variable.

4.17 $r = \frac{\sum (x-\overline{x})(y-\overline{y})}{(n-1)S_x S_y} = \frac{1}{n-1}\sum\left[\left(\frac{x-\overline{x}}{S_x}\right)\left(\frac{y-\overline{y}}{S_y}\right)\right]$

$r = \frac{1}{n-1}\sum (x'y')$

It can be seen that $\sum(x'y') = (n-1)r$. That is, $\sum(x'y')$ is r multiplied by $(n - 1)$.

4.19 The sign of r is determined by the numerator quantity
$\Sigma xy - \dfrac{(\Sigma x)(\Sigma y)}{n}$. Let w be the missing y-value. From the data, n
= 5, Σx = 15, Σy = 10 + w, Σxy = 30 + 5w and
$\Sigma xy - \dfrac{(\Sigma x)(\Sigma y)}{n}$ = (30 + 5w) $- \dfrac{15(10 + w)}{5}$ = 30 + 5w - 30 - 3w = 2w

Since w ≥ 0, $\Sigma xy - \dfrac{(\Sigma x)(\Sigma y)}{n}$ = 2w ≥ 0 and hence r cannot be
negative.

4.21

Body Mass(x)	Time on Diet(y)	(x rank - 8.5)(y rank - 8.5)
1	1	(-7.5)(-7.5) = 56.25
2	3	(-6.5)(-5.5) = 35.75
3	4	(-5.5)(-4.5) = 24.75
4	11	(-4.5)(2.5) = -11.25
5	8	(-3.5)(-.5) = 1.75
6	2	(-2.5)(-6.5) = 16.25
7	7	(-1.5)(-1.5) = 2.25
8	9	(-.5)(.5) = -0.25
9	5	(.5)(-3.5) = -1.75
10	12	(1.5)(3.5) = 5.25
11	6	(2.5)(-2.5) = -6.25
12	15	(3.5)(6.5) = 22.75
13	10	(4.5)(1.5) = 6.75
14	16	(5.5)(7.5) = 41.25
15	13	(6.5)(4.5) = 29.25
16	14	(7.5)(5.5) = 41.25
		264.00

$$r_s = \frac{264.00}{15(16)(17)/12} = \frac{264}{340} = .7765$$

The value for Spearman's rank correlation is somewhat larger than Pearson's correlation coefficient for this data set.

4.23 a) i)

Gang(x)	%Black(y)	(x rank - 4)(y rank - 4)
2	3	(-2)(-1) = 2
4	6	(0)(2) = 0
7	7	(3)(3) = 9
5	5	(1)(1) = 1
6	4	(2)(0) = 0
3	2	(-1)(-2) = 2
1	1	(-3)(-3) = 9
		23

$$r_s = \frac{23}{7(6)(8)/12} = \frac{23}{28} = .821$$

ii) Gang(x)   Income(y)   (x rank - 4)(y rank - 4)

Gang(x)	Income(y)	(x rank - 4)(y rank - 4)
2	7	(-2)(3) = -6
4	6	(0)(2) = 0
7	1	(3)(-3) = -9
5	2	(1)(-2) = -2
6	3	(2)(-1) = -2
3	5	(-1)(1) = -1
1	4	(-3)(0) = 0
		-20

$$r_s = -\frac{20}{28} = -.714$$

iii)

Gang(x)   Poverty(y)   (x rank - 4)(y rank - 4)

Gang(x)	Poverty(y)	(x rank - 4)(y rank - 4)
2	1	(-2)(-3) = 6
4	4	(0)(0) = 0
7	7	(3)(3) = 9
5	5	(1)(1) = 1
6	6	(2)(2) = 4
3	3	(-1)(-1) = 1
1	2	(-3)-2) = 6
		27

$$r_s = \frac{27}{28} = .964$$

b)   Since r_s is largest in absolute value for the percent below poverty level variable, it exhibits the strongest relationship with ganging prevalence.

4.25   Judge 1(x)   Judge 2(y)   (x rank-5)(y rank-5)

Judge 1(x)	Judge 2(y)	(x rank-5)(y rank-5)
7	9	(2)(4) = 8
1	4	(-4)(-1) = 4
3	1	(-2)(-4) = 8
2	3	(-3)(-2) = 6
8	7	(3)(2) = 6
5	5	(0)(0) = 0
9	6	(4)(1) = 4
6	8	(1)(3) = 3
4	2	(-1)(-3) = 3
		42

$$r_s = \frac{42}{9(8)(10)/12} = \frac{42}{60} = .7$$

4.27 a)

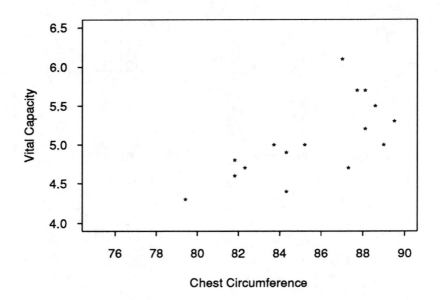

The graph reveals a moderate linear relationship between x and y.

b)

$$b = \frac{6933.48 - \frac{(1368.1)(80.9)}{16}}{117123.85 - \frac{(1368.1)^2}{16}}$$

$$b = \frac{6933.48 - 6917.456}{117123.85 - 116981.101}$$

$$b = \frac{16.024}{142.749} = 0.1123$$

$$a = \frac{80.9}{16} - 0.1123\left(\frac{1368.1}{16}\right)$$

$$a = 5.0563 - 0.1123(85.5063)$$

$$a = 5.0563 - 9.6024 = -4.546$$

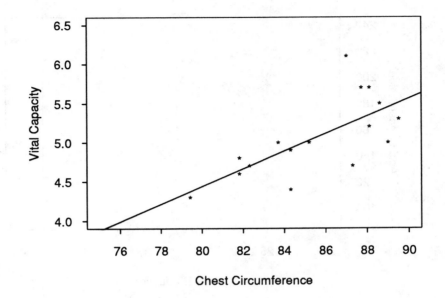

c) The change in vital capacity associated with a 1 cm. increase in chest circumference is .1123.

The change in vital capacity associated with a 10 cm. increase in chest circumference is 10(.1123) = 1.123

d) $\hat{y} = -4.54 + .1123(85) = 5.0055$

e) No; this is shown by the fact that there are two data points in the data set whose x-values are 81.8, but these data points have different y- values.

4.29 a)

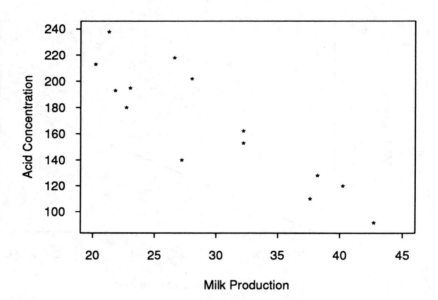

b) $b = \dfrac{-3964.486}{762.012} = -5.2027$

a = 167.43 - (-5.2027)(29.56) = 321.2413

The least squares line is $\hat{y}$ = 321.2413-5.2027x

c)   When x = 25, $\hat{y}$ = 321.2413 - 5.2027(25) = 191.1738
When x = 40, $\hat{y}$ = 321.2413 - 5.2027(40) = 113.1333

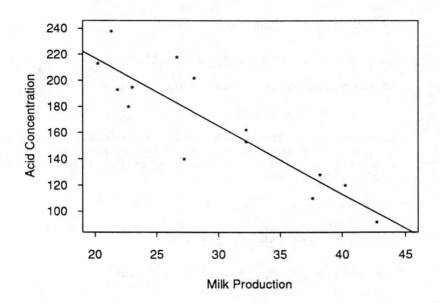

Milk Production

d) $\hat{y}$ = 321.2413-5.2027(30)=165.16

e)   The value 10 is quite far outside of the range over which
      data was collected for milk production. It would be risky to
      predict acid concentration for a cow with milk production of
      10 unless there was a strong reason to suspect that the
      relationship between milk production and acid concentration
      has not changed. If a prediction was made there would be the
      danger of extrapolation present in that prediction.

4.31 a)

ROW	X	Y	X-SQ	XY	Y-SQ
1	25	155	625	3875	24025
2	36	184	1296	6624	33856
3	48	180	2304	8640	32400
4	59	220	3481	12980	48400
5	62	280	3844	17360	78400
6	72	163	5184	11736	26569
7	80	230	6400	18400	52900
8	100	222	10000	22200	49284
9	100	241	10000	24100	58081
10	137	350	18769	47950	122500
Total	719	2225	61903	173865	526415

$$n = 10, \ \overline{x} = \frac{719}{10} = 71.9, \ \overline{y} = \frac{2225}{10} = 222.5$$

$$\Sigma xy - \frac{(\Sigma x)(\Sigma y)}{n} = 173865 - \frac{(719)(2225)}{10} = 13887.5$$

$$\Sigma x^2 - \frac{(\Sigma x)^2}{n} = 61903 - \frac{(719)^2}{10} - 10206.9$$

$$b = \frac{13887.5}{10206.9} = 1.3606$$

$$a = 222.5 - 1.3606(71.9) = 124.6729$$

The equation of the least squares line is

$$\hat{y} = 124.6729 + 1.3606x$$

b)   The value of b is 1.3606. This means that the estimate of the average change in oxygen consumption rate for juvenile steelhead associated with a one-unit increase in plasma cortisol is 1.3606.

c) $\hat{y} = 124.6729 + 1.3606(50) = 192.70$

4.33 a) $n = 15$, $\Sigma x = 82.82$, $\Sigma y = 12545$, $\Sigma x^2 = 459.9784$, $\Sigma y^2 = 12734425$, $\Sigma xy = 67703.9$

$$\Sigma xy - \frac{(\Sigma x)(\Sigma y)}{n} = 67703.9 - \frac{(82.82)(12545)}{15} = -1561.227$$

$$\Sigma x^2 - \frac{(\Sigma x)^2}{n} = 459.9784 - \frac{(82.82)^2}{15} = 2.702$$

$$b = \frac{-1561.227}{2.702} = -577.895$$

$$a = 836.33 - (-577.8953)(5.5213) = 4027.083$$

$$\hat{y} = 4027.083 - 577.895x$$

b)   The b value of -577.895 is the estimate of the average change in myoglobin level associated with a one unit increase in finishing time.

c) $\hat{y} = 4027.083 - 577.895(8) = -596.077$

The least squares equation yields a negative value for the estimated level of myoglobin when the finishing time is 8h. This is clearly unreasonable since myoglobin level cannot be a negative value.

4.35 It is dangerous to use the least squares line to obtain predictions for x-values outside the range of those contained in the sample, because there is no information in the sample about the relationship that exists between x and y for that range of x. The relationship may be the same or it may change substantially. There is no data to support a conclusion either way.

4.37 The denominators of b and of r are always positive numbers. The numerator of b and r is

$$\Sigma(x-\bar{x})(y-\bar{y}) .$$

Since both b and r have the same numerator and positive denominators, they will always have the same sign.

4.39 $r^2 = 1 - \dfrac{5987.16}{17409.60} = 1 - .3439 = .6561$

So 65.61% of the observed variation in age is explained by a linear relationship between percentage of root with transparent dentine for premolars and age.

$s_e^2 = \dfrac{5987.16}{36-2} = \dfrac{5987.16}{34} = 176.0929$

$s_e = \sqrt{176.0929} = 13.27$

The typical amount by which an observation age deviates from the least squares line of percentage of root with transparent dentine and age is 13.27.

4.41   a)

$b = \dfrac{659,010 - \dfrac{(3776)(1773)}{11}}{1365310 - \dfrac{(3776)^2}{11}}$

$b = \dfrac{659010 - 608622.5455}{1365310 - 1296197.8182}$

$b = \dfrac{50387.4545}{69112.1818} = .72907$

$a = \dfrac{1773}{11} - .72907\left(\dfrac{3776}{11}\right) = 161.18182 - 250.26903$

$a = -89.08722$

The least squares equation is $\hat{y} = -89.08722 + .72907x$. For a squawfish whose length is 375, the predicted maximum size is

$-89.08722 + .72907(375)$

$= -89.08722 + .72907(375)$
$= -89.08722 + 273.40125$

$= 184.31403$

The residual at (375, 165) would be
165 - 184.31403 = -19.31403.

b) $SSTo = 323931 - \dfrac{(1773)^2}{11}$
$= 323931 - 285775.3636$
$= 38155.6364$

$SSResid = 323931 - (-89.08722)(1773) -$
$.72907(659010)$

$= 323931 + 157951.64106 - 480464.42070$
$= 1418.22036$

$r^2 = 1 - \dfrac{1418.22036}{38155.6364} = 1 - .03717 = .96283$

4.43 a)

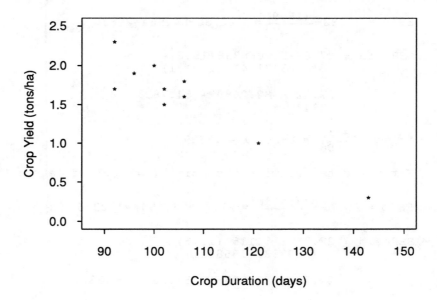

The plot suggests that the least squares line will give fairly accurate predictions. The least squares equation is $\hat{y}$ = 5.20683-.03421x.

b)   The summary statistics for the data remaining after the point (143, .3) is deleted are:

$\Sigma x = 1060-143 = 917$ $\Sigma x^2 = 114514-(143)^2 = 84065$

$\Sigma y = 15.8-.3 = 15.5$ $\Sigma y^2 = 27.82-(.3)^2 = 27.73$

$\Sigma xy = 1601.1 -(143)(.3) = 1558.2$ $n = 9$

$\Sigma x^2 - \dfrac{(\Sigma x)^2}{n} = 94065 - \dfrac{(917)^2}{9}$

$= 94065 - 93432.1111 = 632.8889$

$\Sigma xy - \dfrac{(\Sigma x)(\Sigma y)}{n} = 1558.2 - \dfrac{(917)(15.5)}{9}$
$= 1558.2 - 1579.2778 = -21.0778$

$b = \dfrac{-21.0778}{632.8889} = -.0333$

$a = 1.7222 -(-.0333)(101.8889)=1.7222+3.3930$
$= 5.1151$

The least squares equation with the point deleted is $\hat{y}$ = 5.1151 - .0333x. The deletion of this point does not greatly affect the equation of the line.

c)   For the full data set:

$$SSTo = 27.82 - \frac{(15.8)^2}{10} = 27.82 - 24.964 = 2.856$$

$$SSResid = 27.82 - 5.2068338(15.8) - $$
$$(-.03421541)(1601.1)$$

$$= 27.82 - 82.2680 + 54.7823$$
$$= .3343$$

$$r^2 = 1 - \frac{.3343}{2.856} = 1 - .1171 = .8829$$

For the data set with the point (143, .3) deleted:

$$SSTo = 27.73 - \frac{(15.5)^2}{9} = 27.73 - 26.6944 = 1.0356$$

$$SSResid = 27.73 - 5.1151(15.5) - $$
$$(-.0333)(1558.2)$$

$$= 27.73 - 79.2856 + 51.8881 = .3325$$

$$r^2 = 1 - \frac{.3325}{1.0356} = 1 - .3211 = .6789$$

4.45 a) $$SSResid = 35634 - (-19.669528)(572) -$$
$$(3.284692)(14022)$$
$$= 827.019$$

b)   From the summary statistics given in Exercise 4.30

$$SSTo = 35634 - \frac{(572)^2}{12} = 8368.6667$$
$$r^2 = 1 - \frac{827.019}{8368.6667} = 0.9012$$
Hence, 90.12% of the observed variation in y can be attributed to an approximate linear relationship between x and y.

c) $$SSResid = 35634 - (-19.67)(572) - (3.28)(14022)$$
$$= 893.08$$

The using of the rounded values for a and b in the computation of SSResid results in a substantial difference.

4.47 a,b)

ROW	X	Y	PRED Y	RESIDUAL	RESID SQ
1	-5.0	43	44.5024	-1.5024	2.257
2	-3.2	50	50.6349	-0.6349	0.403
3	-2.2	61	54.0418	6.9582	48.417
4	-1.7	63	55.7452	7.2548	52.632
5	-1.6	47	56.0859	-9.0859	82.554
6	-1.5	57	56.4266	0.5734	0.329
7	-0.9	51	58.4708	-7.4708	55.812
8	0.0	60	61.5370	-1.5370	2.362
9	0.0	67	61.5370	5.4630	29.845
10	1.2	76	65.6253	10.3747	107.635
11	1.6	70	66.9880	3.0120	9.072
12	1.7	51	67.3287	-16.3287	266.627
13	2.8	74	71.0763	2.9237	8.548

SSRESID = 666.493

c) $SSResid = 46940 - (61.536983)(770) -$
    $(3.406908)(-325.8)$
    $= 666.4937$

d) $SSResid = 46940 - (61.5)(770) - (3.4)(-325.8)$
    $= 692.72$
    Yes, rounding has made a substantial difference.

e) $r^2 = 1 - \dfrac{666.4937}{1332.3077} = 0.4997$

Only 49.97% of the total variation in cholesterol level can be explained using a straight line model that relates cholesterol level to percent change in body weight. Looking at the graph of Example 4.9, the relationship seems approximately linear. The data point (1.7, 51) may be part of the reason for a rather low r^2 value.

4.49 a) Whether s_e is small or not depends upon the physical setting of the problem. An s_e of 2 feet when measuring heights of people would be intolerable, while an s_e of 2 feet when measuring distances between planets would be very satisfactory. It is possible for the linear association between x and y to be such that r^2 is large and yet have a value of s_e that would be considered large. Consider the following two data sets:

Set 1		Set 2	
x	y	x	y
5	14	14	5
6	16	16	15
7	17	17	25
8	18	18	35
9	19	19	45
10	21	21	55

For set 1, $r^2 = .981$ and $s_e = .378$.
For set 2, $r^2 = .981$ and $s_e = 2.911$.

Both sets have a large value for r^2, but s_e for data set 2 is 7.7 times larger than s_e for data set 1. Hence, it can be argued that data set 2 has a large r^2 and a large s_e.

b) Now consider the data set

x	5	55	15	45	25	35
y	10.004	10.006	10.007	10.008	10.009	10.010

This data set has $r^2 = .12$ and $s_e = .002266$. So yes, it is possible for a bivariate data set to have both r^2 and s_e small.

c) When r^2 is large and s_e is small, then not only has a large proportion of the total variability in y been explained by the linear association between x and y, but the typical error of prediction is small.

4.51 a)
$$\sum[y-(a+bx)] = \sum[y-(\overline{y}-b\overline{x}+bx)]$$
$$= \sum[(y-\overline{y})-b(x-\overline{x})] = \sum(y-\overline{y})-b\sum(x-\overline{x})$$
$$= 0 + b0 = 0$$

since both $\sum(y-\overline{y})$ and $\sum(x-\overline{x})$ equal 0

b)
$$\frac{\sum(x-\overline{x})(e-\overline{e})}{\sum(x-\overline{x})^2} = \frac{\sum(x-\overline{x})(y-\overline{y})-b(x-\overline{x})}{\sum(x-\overline{x})^2}$$

$$= \frac{\sum(x-\overline{x})(y-\overline{y})}{\sum(x-\overline{x})^2} - b\frac{\sum(x-\overline{x})^2}{\sum(x-\overline{x})^2} = b-b = 0$$

4.53   a)

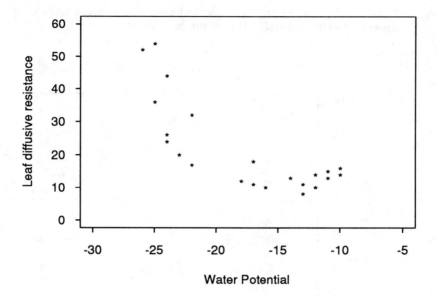

   b)   The plot looks like segment 3 of Figure 4.26. This suggests
        going down the ladder on y or x. One possible transformation
        would be 1/y.

4.55   a)

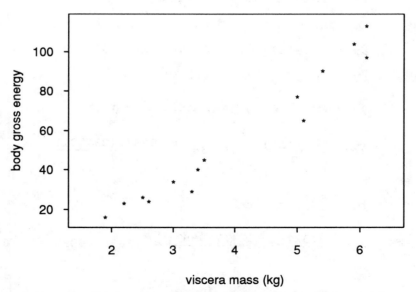

There appears to be curvature (upward) in the graph.

b) $r^2 = 1 - \dfrac{657.27}{15114.93} = 1 - .0435 = .9565$

95.65% of the observed variation in y is explained by the linear relationship between x and y.

c)

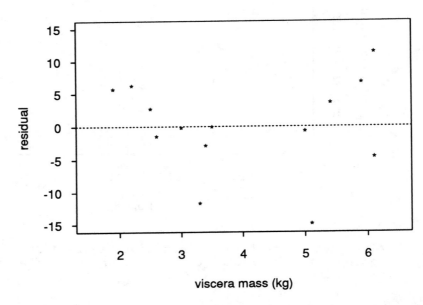

viscera mass (kg)

The residual plot reveals more clearly the curvature in the relationship between x and y.

d)   For the transformed data:

$\Sigma x = 7.9857$, $\Sigma x^2 = 4.9538$, $\Sigma y = 23.321$

$\Sigma y^2 = 39.890$, $\Sigma xy = 13.938$

$\Sigma x^2 - \dfrac{(\Sigma x)^2}{n} = 4.9538 - \dfrac{(7.9857)^2}{14} = 0.3987$

$\Sigma xy - \dfrac{(\Sigma x)(\Sigma y)}{n} = 13.938 - \dfrac{(7.9857)(23.321)}{14}$
$\qquad = 0.6355$

$b = \dfrac{.6355}{.3987} = 1.5940$

$a = \dfrac{23.321}{14} - 1.5940\left(\dfrac{7.9857}{14}\right) = 1.6658 - .9092 = .7565$

So the equation of the least squares line for the transformed data is

$$\hat{y} = .7565 + 1.5940x$$

or

$$\widehat{\log(y)} = .7565 + 1.5940 \log(x)$$

For the transformed data,

$$\text{SSResid} = 39.890 - .7565(23.321) - 1.5940(13.938) = .0294$$

$$\text{SSTo} = 39.890 - \frac{(23.321)^2}{14} = 1.0422$$

$$r^2 = 1 - \frac{.0294}{1.0422} = 1 - .0282 = .9718$$

Since the value of r^2 has increased, a linear relation between the transformed variables seems more reasonable than one between the original variables.

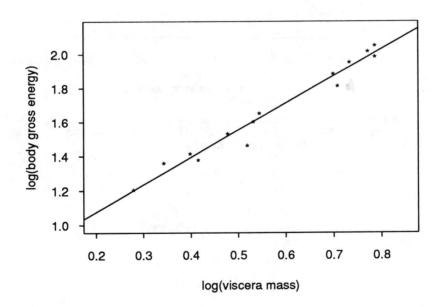

e)   When $x = 5.0$, $\log(x) = \log(5) = .6990$
So the predicted $\log(y)$ when $x = 5$ is

$$\hat{y} = .7565 + 1.5940(.699) = 1.8707$$

Taking the antilog of 1.8707 yields a predicted value of

$$\hat{y} = 10^{1.8707} = 74.2506$$

4.57 a) The correlation between x and y is -.717

b) The correlation between $\sqrt{x}$ and $\sqrt{y}$ is -.835. Since this correlation is larger in absolute value than the correlation of part (a), the transformation appears successful in straightening the plot.

4.59 a)

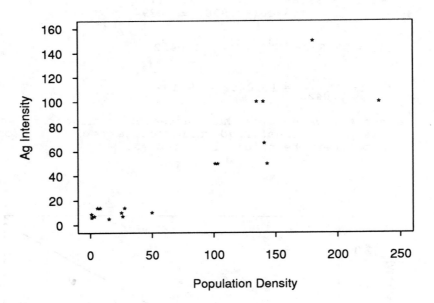

The plot does appear to have a positive slope, so the scatter plot is compatible with the "positive association" statement made in the paper.

b)

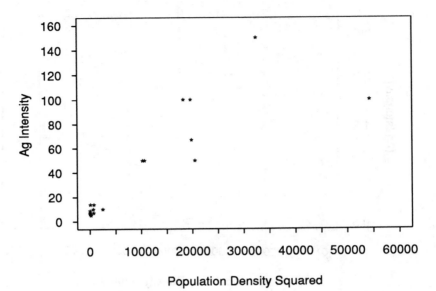

This transformation does straighten the plot, but it also appears that the variability of y increases as x increases.

c)

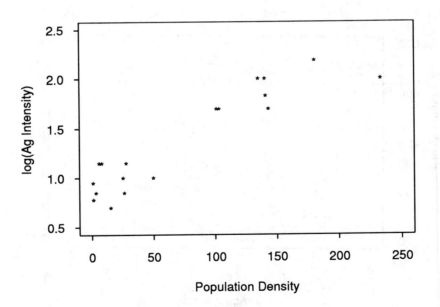

The plot appears to be as straight as the plot in (b), and has the desirable property that the variability in y appears to be constant regardless of the value of x.

d)

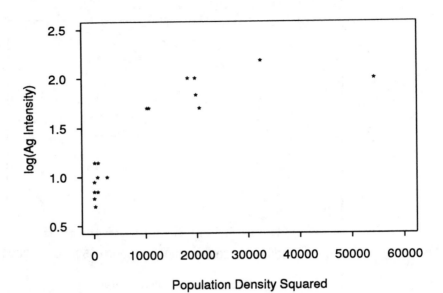

This plot has curvature opposite of the plot in (a), suggesting that this transformation has taken us too far along the ladder.

4.61

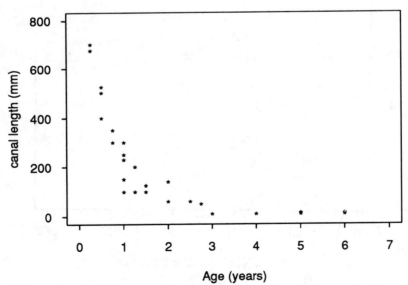

The relationship between age and canal length is not linear, but curvilinear. Transforming to 1/x produces a scatterplot that is much straighter than the plot above.

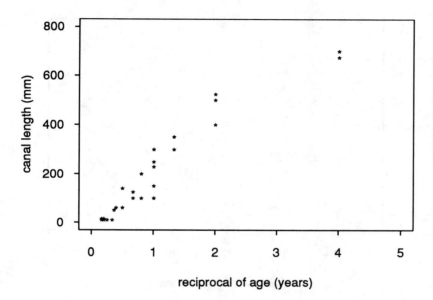

4.63 a)

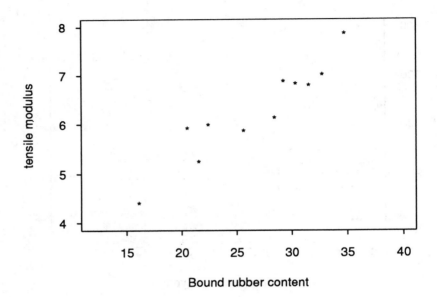

The scatterplot above suggests the plausibility of a linear relationship between bound rubber content and tensile modulus.

b)   The summary values are: n = 11, Σx = 292.9, Σx^2 = 8141.8, Σy = 69.03, Σy^2 = 442.19, Σxy = 1890.2

$$\Sigma xy - \frac{(\Sigma x)(\Sigma y)}{n} = 52.1194$$

$$\Sigma x^2 - \frac{(\Sigma x)^2}{n} = 342.672$$

$$\Sigma y^2 - \frac{(\Sigma y)^2}{n} = 8.9954$$

b = 52.1194/342.672 = .1521
a = 6.2755 - (.1521)(26.6273) = 2.2255
The equation of the estimated regression line is

$$\hat{y} = 2.2255 + .1521x$$

c)   From the computations in part (b),

$$r = \frac{52.1194}{\sqrt{(342.672)(8.9954)}} = 0.93875, \text{ and}$$

$$r^2 = (.93875)^2 = .881$$

Since $r^2 = .881$, this means that 88.1% of the total variability in bound rubber content can be explained using a simple linear regression model.

4.65 a)   y = stride rate   x = speed

The summary values are: n = 11, $\Sigma x = 205.4$, $\Sigma x^2 = 3880.08$, $\Sigma y = 35.16$, $\Sigma y^2 = 112.681$, $\Sigma xy = 660.13$

$$\overline{x} = \frac{205.4}{11} = 18.6727, \quad \overline{y} = \frac{35.16}{11} = 3.1964$$

$$\Sigma xy - \frac{(\Sigma x)(\Sigma y)}{n} = 660.13 - \frac{(205.4)(35.16)}{11} = 3.5969$$

$$\Sigma x^2 - \frac{(\Sigma x)^2}{n} = 3880.08 - \frac{(205.4)^2}{11} = 44.7018$$

$$\Sigma y^2 = \frac{(\Sigma y)^2}{n} = 112.681 - \frac{(35.16)^2}{11} = .2969$$

$$b = \frac{3.5969}{44.7018} = .0805$$
$$a = 3.1964 - (.0805)(18.6727) = 1.6932$$

The equation of the least squares line for predicting stride rate from speed is

$$\hat{y} = 1.6932 + .0805x$$

b)   x = stride rate   y = speed
The summary values are: n = 11, $\Sigma x = 35.16$, $\Sigma x^2 = 112.681$, $\Sigma y = 205.4$, $\Sigma y^2 = 3880.08$, $\Sigma xy = 660.13$

$$\overline{x} = \frac{35.16}{11} = 3.1964, \quad \overline{y} = \frac{205.4}{11} = 18.6727,$$

$$\Sigma xy - \frac{(\Sigma x)(\Sigma y)}{n} = 660.13 - \frac{(35.16)(205.4)}{11} = 3.5969$$

$$\Sigma x^2 - \frac{(\Sigma x)^2}{n} = 112.681 - \frac{(35.16)^2}{11} = .2969$$

$$\Sigma y^2 - \frac{(\Sigma y)^2}{n} = 3880.08 - \frac{(205.4)^2}{11} = 44.7018$$

$$b = \frac{3.5969}{.2969} = 12.1149$$

$a = 18.6727 - (12.1149)(3.1964) = -20.0514$
The equation of the least squares line for predicting speed from stride rate is

$$\hat{y} = -20.0514 + 12.1149x$$

c)   The coefficient of determination for the "stride rate on speed" regression is

$$r^2 = \frac{(3.5969)^2}{(44.7018)(.2969)} = .9748$$

The coefficient of determination for the "speed on stride rate" regression is

$$r^2 = \frac{(3.5969)^2}{(.2969)(44.7018)} = .9748$$

The relationship between the two r^2's is that they are equal in value.

4.67   a)   The data demonstrates unequivocally that quit rate is determined at least in part by factors other than wages because there are two data points (8.83, 1.4) and (8.80, 2.0) which have nearly identical x-values but substantially different y-values.

b)

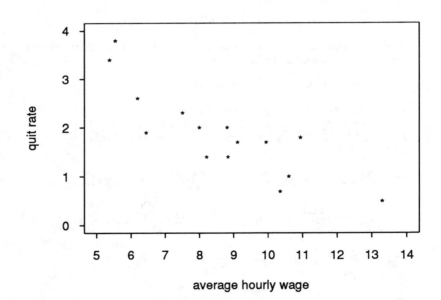

The scatterplot suggests that the relationship between quit rate and average hourly wage might be adequately modeled by a linear equation. However, there is also a hint of curvature to the plot.

c) $\sum xy - \frac{(\sum x)(\sum y)}{n} = 218.806 - \frac{(129.05)(28.2)}{15} = -23.8080$

$\sum x^2 - \frac{(\sum x)^2}{n} = 1178.9601 - \frac{(129.05)^2}{15} = 68.6999$

$$\sum y^2 - \frac{(\sum y)^2}{n} = 64.34 - \frac{(28.2)^2}{15} = 11.3240$$

$$b = \frac{-23.8080}{68.6999} = -.3466$$

$$a = \frac{28.2}{15} - (-.3466)\left(\frac{129.05}{15}\right) = 4.8615$$

$$\hat{y} = 4.8615 - .3466x$$

d) $\hat{y} = 4.8615 - .3466(7.50) = 2.2620$
The residual for the data point (7.5, 2.3) is
$(y - \hat{y}) = 2.3 - 2.2620 = 0.0380$

e) $SSResid = 64.34 - 4.8615(28.2) - (-.3466)(218.806)$
                $= 64.34 - 137.0943 + 75.8273$
                $= 3.0730$

$$r^2 = 1 - \frac{3.0730}{11.3240} = 1 - .2714 = .7286$$

4.69 $\sum x^2 - \frac{(\sum x)^2}{n} = 2.034, \quad s_x = \sqrt{2.034/6} = .5822$

$$\sum y^2 - \frac{(\sum y)^2}{n} = .00428, \quad s_y = \sqrt{.00428/6} = .0267$$

$$\sum xy - \frac{(\sum x)(\sum y)}{n} = .012286$$

$$r = \frac{.012286}{6(.5822)(.0267)} = .132$$

This value of r indicates a weak positive linear relationship.

4.71   a)

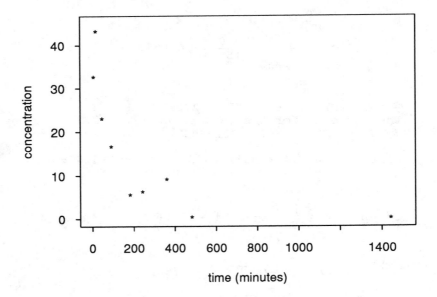

b)   Based on the plot in (a) and figure 4.26 a transformation
        going down the ladder on x or y is suggested. The
        transformation log(time) will produce a reasonably straight
        plot.

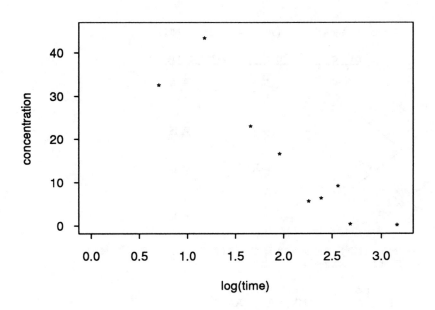

4.73 a) For the first equation, the left side is

$$n\left[\frac{\Sigma y}{n} - \frac{b\Sigma x}{n}\right] + b\Sigma x = \frac{n\Sigma y}{n} - \frac{bn\Sigma x}{n} + b\Sigma x =$$

$$\Sigma y - b\Sigma x + b\Sigma x = \Sigma y$$

For the second equation, the left side is

$$(\Sigma x)(\overline{y} - b\overline{x}) + (\Sigma x^2)b = (\Sigma x)\overline{y} - b(\Sigma x)\overline{x} + (\Sigma x^2)b$$

$$= n\overline{xy} - nb\overline{x}^2 + (\Sigma x^2)b = n\overline{xy} + b[\Sigma x^2 - n\overline{x}^2]$$

$$= n\overline{xy} + \frac{\Sigma xy - n\overline{xy}}{\Sigma x^2 - n\overline{x}^2}(\Sigma x^2 - n\overline{x}^2)$$

$$= n\overline{xy} + \Sigma xy - n\overline{xy} = \Sigma xy$$

b) $\Sigma(y - \hat{y}) = \Sigma[y - (a + bx)] = \Sigma y - (na + b\Sigma x) = \Sigma y - \Sigma y = 0$

5.1 a)   Sample Space = {AA, AM, MA, MM}

b)   <u>1st car</u>   <u>2nd car</u>   <u>Outcome</u>

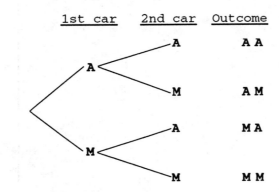

c)   B = {MA, AM, AA}; C = {AM, MA}; D = {MM}
D is a simple event.

d)   (B and C) = {AM, MA};
(B or C) = {MA, AM, AA}

5.3 a)   <u>First book</u>   <u>Second book</u>   <u>Outcome</u>
      selected   selected

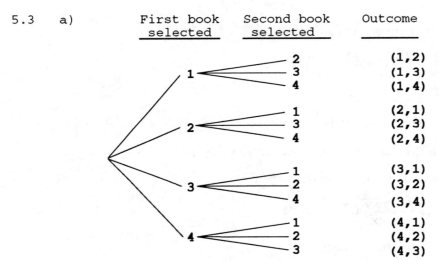

b)   A = {(1,2),(1,4),(2,1),(2,3),(2,4),(3,2),(3,4),
(4,1),(4,2),(4,3)}

c)   B = {(1,3),(1,4),(2,3),(2,4),(3,1),(3,2),(4,1),(4,2)}

5.5 a)   N, DN, DDN, DDDN, DDDDN

b)   There are countably infinitely many.

c)   E = {DN, DDDN, DDDDDN, ...}

5.7 a)
```
(1,1,1)     (2,1,1)     (3,1,1)
(1,1,2)     (2,1,2)     (3,1,2)
(1,1,3)     (2,1,3)     (3,1,3)
(1,2,1)     (2,2,1)     (3,2,1)
(1,2,2)     (2,2,2)     (3,2,2)
(1,2,3)     (2,2,3)     (3,2,3)
(1,3,1)     (2,3,1)     (3,3,1)
(1,3,2)     (2,3,2)     (3,3,2)
(1,3,3)     (2,3,3)     (3,3,3)
```

b) A = {(1,1,1),(2,2,2),(3,3,3)}

c) B = {(1,2,3),(1,3,2),(2,1,3),(2,3,1),(3,1,2),
        (3,2,1)}

d) C = {(1,1,1),(1,1,3),(1,3,1),(1,3,3),(3,1,1),
        (3,1,3),(3,3,1),(3,3,3)}

e) (not B) = {(1,1,1),(1,1,2),(1,1,3),(1,2,1),
              (1,2,2),(1,3,1),(1,3,3),(2,1,1),
              (2,1,2),(2,2,1),(2,2,2),(2,2,3),
              (2,3,2),(2,3,3),(3,1,1),(3,1,3),
              (3,2,2),(3,2,3),(3,3,1),(3,3,2),
              (3,3,3)}

    (not C) = {(1,1,2),(1,2,1),(1,2,2),(1,2,3),
              (1,3,2),(2,1,1),(2,1,2),(2,1,3),
              (2,2,1),(2,2,2),(2,2,3),(2,3,1),
              (2,3,2),(2,3,3),(3,1,2),(3,2,1),
              (3,2,2),(3,2,3),(3,3,2)}

    (A or B) = {(1,1,1),(2,2,2),(3,3,3),(1,2,3),
               (1,3,2),(2,1,3),(2,3,1),(3,1,2),
               (3,2,1)}

    (A and B) = empty set
    (A and C) = {(1,1,1),(3,3,3)}

5.9 a)

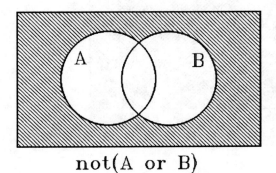

not(A or B)

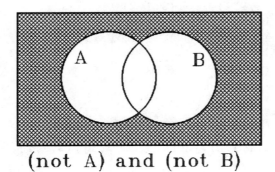

(not A) and (not B)

The two shaded regions are identical.

b)

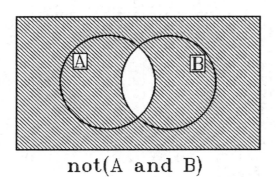

not(A and B)

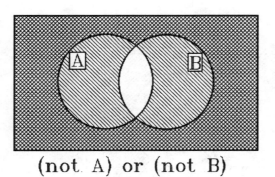

(not A) or (not B)

The shaded area of not(A and B) is identical to the area of (not A) or (not B) covered by lines running from lower left to upper right or by upper left to lower right.

5.11 a) $P(\text{twins}) = \dfrac{500,000}{42,005,100} = .0119$

b) $P(\text{quadruplets}) = \dfrac{100}{42,005,100} = .00000238$

c) P(more than a single child)
$= \dfrac{505,100}{42,005,100} = .012$

5.13 a) P(request is for one of the 3 B's)
$= .05 + .26 + .09 = .40$

b) P(not for one of the two S's)
$= 1 - P(\text{for one of the two S's})$
$= 1 - (.12 + .07) = .81$

c) P(request for a composer who wrote at least one symphony)
$= 1 - P(\text{composer did not write a symphony})$
$= 1 - (.05 + .01) = .94$

d) P(request is for a piece by one of the main characters in the movie *Amadeus*)
$= P(\text{Mozart}) = .21$

5.15 a) There are 52 simple events.

b) $\dfrac{1}{52}$

c) $P(\text{heart}) = \dfrac{13}{52} = \dfrac{1}{4}$

$P(\text{face card}) = \dfrac{12}{52} = \dfrac{3}{13}$

d) $P(\text{both a face card and a heart}) = \dfrac{3}{52}$

e) $P(A \text{ or } B) = P(A) + P(B) - P(A \text{ and } B)$
$= \dfrac{12}{52} + \dfrac{13}{52} - \dfrac{3}{52} = \dfrac{22}{52} = \dfrac{11}{26}$

5.17 a) The outcome (2,4,3,1) is given in the problem. The other 23 possible outcomes are:

(1,2,3,4)	4	(3,1,2,4)	1
(1,2,4,3)	2	(3,1,4,2)	0
(1,4,2,3)	1	(3,4,1,2)	0
(1,4,3,2)	2	(3,4,2,1)	0
(1,3,4,2)	1	(3,2,1,4)	2
(1,3,2,4)	2	(3,2,4,1)	1
(2,1,3,4)	2	(4,1,2,3)	0
(2,1,4,3)	0	(4,1,3,2)	1
(2,4,1,3)	0	(4,2,3,1)	2
(2,3,1,4)	1	(4,2,1,3)	1
(2,3,4,1)	0	(4,3,1,2)	0
		(4,3,2,1)	0

b)   (1,2,4,3), (1,4,3,2), (1,3,2,4), (2,1,3,4), (3,2,1,4), (4,2,3,1)

P(exactly two of the books are returned to their correct owners) $= \frac{6}{24} = .2500$

c)   P(exactly one receives his/her book)

$= \frac{8}{24} = .3333$

d)   P(exactly three receive their own book) = 0

e)   P(at least two of the four students receive their own books)

$= \frac{(6 + 1)}{24} = \frac{7}{24} = .2917$

5.19   a)   The ten simple events are (B,C) (B,M) (B,P) (B,S) (C,M) (C,P) (C,S) (M,P) (M,S) (P,S).

b)   Each would be equally likely with probability .1.

c)   P(committee includes the statistics representative)

$= P[(B,S),(C,S),(M,S),(P,S)] = .4$

d)   P(both members are from "laboratory resources")

$= P[(B,C),(B,P),(C,P)] = .3$

5.21   The simple events are (1,2) (1,3) (1,4) (1,5) (2,3) (2,4) (2,5) (3,4) (3,5) (4,5).

a)   P(both are first printings) $= P[(1,2)] = \frac{1}{10}$

b)   P(both are second printings) =

$P[(3,4),(3,5)$ or $(4,5)] = \frac{3}{10}$

c)   P(at least one is a first printing)

$= 1 - P(\text{neither is a first printing})$

$= 1 - \frac{3}{20} = \frac{7}{10}$

d)   P(copies are of different printings) $= \frac{6}{10}$

5.23   a) $P(E|F) = \dfrac{P(E \text{ and } F)}{P(F)} = \dfrac{.54}{.6} = .9$

       b) $P(F|E) = \dfrac{P(E \text{ and } F)}{P(E)} = \dfrac{.54}{.7} = .771$

       c)   E and F are not independent because $P(E|F) \neq P(E)$.

5.25   a) $P(TP|Oil) = \dfrac{P(TP \text{ and } Oil)}{P(Oil)} = \dfrac{.008}{.10} = .08$

       b) $P(Oil|TP) = \dfrac{P(TP \text{ and } Oil)}{P(TP)} = \dfrac{.008}{.04} = .2$

       c)   The events "checks oil pressure" and "checks tire pressure" are not independent, because $P(TP|Oil) \neq P(TP)$.

5.27   a) $P(A) = .32 + .27 + .18 = .77$
                 $P(FD) = .27 + .04 = .31$

       b) $P(A|FD) = \dfrac{P(A \text{ and } FD)}{P(FD)} = \dfrac{.27}{.31} = .871$

       c) $P(M|\text{not } HB) = \dfrac{P(M \text{ and not } HB)}{P(\text{not } HB)} =$

           $\dfrac{(.08+.04)}{(.40+.31)} = \dfrac{.12}{.71} = .169$

       The conditional probability of manual transmission given not a hatchback is smaller than the unconditional probability of a manual transmission.

5.29   a) $P(E) = \dfrac{6}{10}$

       b) $P(F|E) = \dfrac{5}{9}$

       c) $P(E \text{ and } F) = P(E)P(F|E) = \left(\dfrac{6}{10}\right)\left(\dfrac{5}{9}\right) = \dfrac{30}{90} = \dfrac{1}{3}$

5.31   a) $P(\text{1-2 subsystem works}) = .9(.9) = .81$
                 $P(\text{3-4 subsystem works}) = .9(.9) = .81$

       b) $P(\text{both subsystems function}) = .81(.81) = .6561$

       c) $P(\text{system functions})$
           $= P(\text{1-2 subsystem works}) + P(\text{3-4 subsystem works}) - P(\text{both subsystems work})$
           $= .81 + .81 - .6561 = .9639$

5.33   a) $P(\text{all three flights full}) = .6(.5)(.4) = 0.12$
                 $P(\text{at least one flight not full}) = 1 - .12 = .88$

b)   P(only the New York flight is full) = .6(.5)(.6) = 0.18
        P(exactly one flight full) =
        .6(.5)(.6) + (.4)(.5)(.6) + (.4)(.5)(.4) =
        .18 + .12 + .08 = .38

5.35 $P(B_1$ and S) = .4(.3) = .12 $P(B_2$ and S) = .6(.3) = .18
        $P(B_1$ and M) = .4(.5) = .20 $P(B_2$ and M) = .6(.5) = .30
        $P(B_1$ and L) = .4(.2) = .08 $P(B_2$ and L) = .6(.2) = .12

5.37   a)   P(E or F) = P(E) + P(F) - P(E and F) = .4 + .3 - .15 = .55

       b)   P(neither E nor F) = 1 - P(E or F) = 1 - .55 = .45

       c)   P(exactly one) = P(E or F) - P(E and F) = .55 - .15 = .40

       d)   P(must stop at first light only) =
             P(E) - P(E and F) = .4 - .15 = .25

5.39   a)   P(medium auto and high homeowner) = .10

       b)   P(low auto) = .04 + .06 + .05 + .03 = .18
             P(low homeowner) = .06 + .10 + .03 = .19

       c)   P(in same category for both types of insurance) = P(L,L) +
             P(M,M) + P(H,H) = .06 + .20 + .15 = .41

       d)   P(not in same category for both types)
             = 1 - P(in same category for both types) = 1 - .41 = .59

       e)   P(an individual has at least one low deductible) = P(low
             auto) + P(low homeowner) - P(both low) = .18 + .19 - .06 =
             .31

       f)   P(neither deductible level is low)
             = 1 - P(at least one is low) = 1 - .31 = .69

5.41   a)

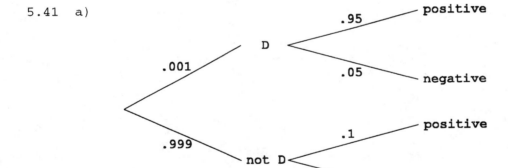

       b)   P(has disease and positive result) = .001(.95) = .00095

       c)   P(positive result) = P(has disease and positive)
             + P(not disease and positive)
             = .001(.95) + .999(.1) = .00095 + .0999 =
             .10085

       d)   P(has disease|positive result)
             = $\frac{.00095}{.10085}$ = .00942

5.43   P(E$_1$ and L) = P(E$_1$)P(L|E$_1$) = .4(.02) = .008

5.45 There are 27(41) = 1,107 ways to listen first to a Mozart piano concerto and then to a symphony.

5.47 a) $\qquad$ 6(5)(4) = 120

b) $\qquad \binom{15}{3} = \dfrac{15!}{3!12!} = \dfrac{15(14)(13)}{3(2)(1)} = \dfrac{2730}{6} = 455$

c) $\qquad \dfrac{120}{455} = 0.2637$

5.49 5(8)(6) = 240 nights.

5.51 a) $\qquad P_{15,\,9} = 15(14)(13)(12)(11)(10)(9)(8)(7)$
$\qquad\qquad = 1,816,214,400$

b) $\qquad$ If A, B, C and D are left on the bench then the starting nine must be selected from the remaining eleven players. This can be done in $P_{11,\,9} = 11(10)(9)(8)(7)(6)(5)(4)(3) = 19,958,400$ ways. Hence the probability that players A, B, C and D will be left on the bench is

$$\dfrac{19,958,400}{1,816,214,400} = 0.011.$$

5.53   P(at least two disks must be examined)
         = 1 - P(only one examined)
         = 1 - P(first examined disk is blank)
         = $1 - \dfrac{15}{25} = \dfrac{10}{25} = .4$

5.55   The possible peer review committees are

       (3,6)   (3.7)   (3,10) $\underline{(3,14)}$ (6,7)

       $\underline{(6,10)}$ $\underline{(6,14)}$ $\underline{(7,10)}$ $\underline{(7,14)}$ $\underline{(10,14)}$.

       The underlined committees have a total of at least 15 years
       experience. The probability of selecting a committee with
       representatives that have a total of at least 15 years of teaching
       experience is 6/10 = .6.

5.57   P(a randomly selected subscriber reads at least one of these three
       columns) = .14 + .23 + .37 - .08 - .09 -.13 + .05 = 0.49

5.59 R_1 and R_2 are not independent events since $P(R_2|R_1)$ = 1139/2516 =
       .4527 which is not equal to $P(R_2)$ which is .4529. However, these
       values do not differ much and thus one could say they are
       independent for practical purposes.

5.61   a)   P(medium|short-sleeve plaid)

                 $\dfrac{\text{P(medium and short-sleeve plaid)}}{\text{P(short-sleeve plaid)}}$

                 $= \dfrac{.08}{(.04 + .08 + .03)} = \dfrac{.08}{.15} = .533$

       b)   P(medium|short-sleeve plaid)

                 $\dfrac{\text{P(short-sleeve and medium plaid)}}{\text{P(medium plaid)}}$

                 $= \dfrac{.08}{(.08 + .10)} = \dfrac{.08}{.18} = .444$

            P(long-sleeve|medium plaid)

                 $\dfrac{\text{P(long-sleeve and medium plaid)}}{\text{P(medium plaid)}}$

                 $= \dfrac{.10}{(.08 + .10)} = \dfrac{.10}{.18} = .556$

5.63   a)   Yes, if the slip that reads "win prizes 1, 2 and 3" is
            drawn. $P(E_1 \text{ and } E_2)$ = 1/4, $P(E_1)$ = 2/4, and $P(E_2)$ = 2/4.
            Since $P(E_1 \text{ and } E_2) = P(E_1)P(E_2)$, the events E_1 and E_2 are
            independent.

       b) $P(E_1 \text{ and } E_3)$ = 1/4, $P(E_1)$ = 1/2, and $P(F_3)$ = 1/2
            Since $P(E_1 \text{ and } E_3) = P(E_1)P(E_3)$, the events E_1 and E_3 are
            independent.
            $P(E_2 \text{ and } E_3)$ = 1/4, $P(E_2)$ = 1/2, and $P(E_3)$ = 1/2
            Since $P(E_2 \text{ and } E_3) = P(E_2)P(E_3)$, the events E_2 and E_3 are
            independent.

c) $P(E_1 \text{ and } E_2 \text{ and } E_3) = \dfrac{1}{4}$

$P(E_1)(E_2)(E_3) = \dfrac{1}{2}\left(\dfrac{1}{2}\right)\left(\dfrac{1}{2}\right) = \dfrac{1}{8}$

So $P(E_1 \text{ and } E_2 \text{ and } E_3) \neq P(E_1)P(E_2)P(E_3)$

5.65 a)   P(fills and pays with a credit card)
         = P(fills)P(pays with credit card|fills)
         = .4(.8) = .32

      b)   P(all three fill their tanks and pay by credit card)
         = $(.32)^3$
         = .032768

5.67 a) $P(A_1) = \dfrac{3}{15} = .2$

$P(A_2|A_1) = \dfrac{2}{14} = .143$

$P(A_1 \text{ and } A_2)$ = P(both applications selected for processing are for 36 months) = $P(A_1)P(A_2|A_1)$

$= \left(\dfrac{3}{15}\right)\left(\dfrac{2}{14}\right) = \dfrac{1}{35} = .0286$

      b)   P(both are the same duration) = P(both are for 36 months) + P(both are for 48 months) + P(both are for 60 months)

$= \left(\dfrac{3}{15}\right)\left(\dfrac{2}{14}\right) + \left(\dfrac{5}{15}\right)\left(\dfrac{4}{14}\right) + \left(\dfrac{7}{15}\right)\left(\dfrac{6}{14}\right)$

$= \dfrac{(6 + 20 + 42)}{210} = \dfrac{68}{210} = .324$

5.69 a) $P(B_1 \text{ and } S) = P(S)P(B_1|S) = .2(.5) = .10$
$P(T) = 1 - P(S) = .8$
$P(B_2 \text{ and } T) = P(T) - P(B_1 \text{ and } T) - P(B_3 \text{ and } T)$
       $= .8 - .35 - .26 = .19$
$P(B_2 \text{ and } T) = P(B_2)P(T|B_2)$
    $.19 = P(B_2)(.76)$

$P(B_2) = \dfrac{.19}{.76} = .25$

$P(B_2 \text{ and } S) = P(B_2) - P(B_2 \text{ and } T) = .25 - .19 = .06$
$P(B_3 \text{ and } S) = P(S) - P(B_1 \text{ and } S) - P(B_2 \text{ and } S)$
       $= .2 - .10 - .06 = .04$

      b) $P(\text{not } B_1|T) = 1 - P(B_1|T) = 1 - \dfrac{P(B_1 \text{ and } T)}{P(T)}$

$= .1 - \dfrac{.35}{.80} = \dfrac{.45}{.80} = .5625$

5.71 a)   P(all three are hearts) = $P(H_1)P(H_2|H_1)P(H_3|H_1 \text{ and } H_2)$
$= \left(\dfrac{13}{52}\right)\left(\dfrac{12}{51}\right)\left(\dfrac{11}{50}\right) = .0129$

b)   P(all three cards are from the same suit)
      = P(all three are hearts) +
        P(all three are spades) +
        P(all three are clubs) +
        P(all three are diamonds)
      = 4(.0129) = .0518

c)   P(all five are hearts)
$$= \left(\frac{13}{52}\right)\left(\frac{12}{51}\right)\left(\frac{11}{50}\right)\left(\frac{10}{49}\right)\left(\frac{9}{48}\right) = .000495$$

5.73   Let S denote the bit received is passed on with no reversal and F denote the bit received is passed on with a reversal.

a) $P(S_1 \text{ and } S_2 \text{ and } S_3) = P(S_1)(S_2)(S_3) = .8(.8)(.8) = .512$

b)   P(1 received if a 1 is sent)
      = P(no reversals) + P(2 reversals)
      $= P(S_1 \text{ and } S_2 \text{ and } S_3) + P(S_1 \text{ and } F_2 \text{ and } F_3) +$
        $P(F_1 \text{ and } F_2 \text{ and } S_3) + P(F_1 \text{ and } S_2 \text{ and } F_2)$
      = .8(.8)(.8) + (.8)(.2)(.2) + (.2)(.2)(.8)
      + (.2)(.8)(.2)
      = .512 + 3(.032) = .512 + .096 = .608

5.75   P(all five are of the same brand)
    = P(all five are Penn) + P(all five are Wilson)
$$= \frac{\binom{12}{5} + \binom{8}{5}}{\binom{20}{5}} = \frac{792 + 56}{15,504} = \frac{848}{15,504} = 0.0547$$

5.77   P(B) = P(A) + P[(not A) and B]
This says that P(B) must be at least as large as P(A).

CHAPTER 6
RANDOM VARIABLES AND
DISCRETE PROBABILITY DISTRIBUTIONS

Section 6.1

6.1 a)   discrete

     b)   continuous

     c)   discrete

     d)   discrete

     e)   continuous

6.3 Possible y values are the positive integers.

Outcomes	y-value
S	1
LLS	3
LRS	3
LS	2
RS	2

6.5 y is a continuous variable with values from 0 to 100 ft.

6.7

Outcomes	x-values	y-values	z-values	w-values
(1,2)	3	-1	1	0
(1,3)	4	-2	0	0
(1,4)	5	-3	1	1
(2,3)	5	-1	1	0
(2,4)	6	-2	2	1
(3,4)	7	-1	1	1

6.9   a) $P(x \leq 100) = .05 + .10 + .12 + .14 + .24 + .17$
           $= .82$

      b) $P(x > 100) = 1 - P(x \leq 100) = 1 - .82 = .18$

      c) $P(x \leq 99) = .05 + .10 + .12 + .14 + .24 = .65$
           $P(x \leq 97) = .05 + .10 + .12 = .27$

6.11 The results of the 50 observations on x are:

      0,0,1,1,1,1,1,2,1,1,1,1,0,1,0,1,0,2,1,1,0,1,1,1,1
      1,1,0,1,1,0,1,1,1,0,1,1,1,1,2,0,1,0,1,0,0,1,1,1,0

Number of Defects	Frequency	Rel. Freq.
0	14	.28
1	33	.66
2	3	.06
		1.00

6.13 a)

Outcomes	Values of x	Probability	
SSSS	4	(.2)(.2)(.2)(.2)	= .0016
SSSF	3	(.2)(.2)(.2)(.8)	= .0064
SSFS	3	(.2)(.2)(.8)(.2)	= .0064
SFSS	3	(.2)(.8)(.2)(.2)	= .0064
FSSS	3	(.8)(.2)(.2)(.2)	= .0064
SSFF	2	(.2)(.2)(.8)(.8)	= .0256
SFSF	2	(.2)(.8)(.2)(.8)	= .0256
FSSF	2	(.8)(.2)(.2)(.8)	= .0256
SFFS	2	(.2)(.8)(.8)(.2)	= .0256
FSFS	2	(.8)(.2)(.8)(.2)	= .0256
FFSS	2	(.8)(.8)(.2)(.2)	= .0256
SFFF	1	(.2)(.8)(.8)(.8)	= .1024
FSFF	1	(.8)(.2)(.8)(.8)	= .1024
FFSF	1	(.8)(.8)(.2)(.8)	= .1024
FFFS	1	(.8)(.8)(.8)(.2)	= .1024
FFFF	0	(.8)(.8)(.8)(.8)	= .4096

x	0	1	2	3	4
p(x)	.4096	.4096	.1536	.0256	.0016

      b)   The most likely values for x are 0 and 1. Each has a
           probability of .4096.

      c) $P(x \geq 2) = .1536 + .0256 + .0016 = .1808$

6.15

Outcomes (slips)	Outcomes ($)	w-values
(1,2)	(1,1)	1
(1,3)	(1,1)	1
(1,4)	(1,10)	10
(1,5)	(1,25)	25
(2,3)	(1,1)	1
(2,4)	(1,10)	10
(2,5)	(1,25)	25
(3,4)	(1,10)	10
(3,5)	(1,25)	25
(4,5)	(10,25)	25

w-values	1	10	25
p(w)	.3	.3	.4

6.17

Outcomes	x-value	Probability
(1,1)	1	$\frac{1}{36}$
(1,2)	2	$\frac{1}{36}$
(1,3)	3	$\frac{1}{36}$
(1,4)	4	$\frac{1}{36}$
(1,5)	5	$\frac{1}{36}$
(1,6)	6	$\frac{1}{36}$
(2,1)	1	$\frac{1}{36}$
(2,2)	2	$\frac{1}{36}$
(2,3)	3	$\frac{1}{36}$
(2,4)	4	$\frac{1}{36}$
(2,5)	5	$\frac{1}{36}$
(2,6)	6	$\frac{1}{36}$
(3,1)	1	$\frac{1}{36}$
(3,2)	2	$\frac{1}{36}$
(3,3)	3	$\frac{1}{36}$
(3,4)	4	$\frac{1}{36}$

(3,5)	5	$\frac{1}{36}$
(3,6)	6	$\frac{1}{36}$
4	4	$\frac{1}{6}$
5	5	$\frac{1}{6}$
6	6	$\frac{1}{6}$

x-value	1	2	3	4	5	6
p(x)	$\frac{1}{12}$	$\frac{1}{12}$	$\frac{1}{12}$	$\frac{1}{4}$	$\frac{1}{4}$	$\frac{1}{4}$

6.19

Outcomes	x-value	Probability
(W,W)	0	(.4)(.4) = .16
(W,T)	1	(.4)(.3) = .12
(W,F)	2	(.4)(.2) = .08
(W,S)	3	(.4)(.1) = .04
(T,W)	1	(.3)(.4) = .12
(T,T)	1	(.3)(.3) = .09
(T,F)	2	(.3)(.2) = .06
(T,S)	3	(.3)(.1) = .03
(F,W)	2	(.2)(.4) = .08
(F,T)	2	(.2)(.3) = .06
(F,F)	2	(.2)(.2) = .04
(F,S)	3	(.2)(.1) = .02
(S,W)	3	(.1)(.4) = .04
(S,T)	3	(.1)(.3) = .03
(S,F)	3	(.1)(.2) = .02
(S,S)	3	(.1)(.1) = .01

y-value	0	1	2	3
p(y)	.16	.33	.32	.19

6.21 a) σ^2 = $(1 - 4.12)^2(.05) + (2 - 4.12)^2(.10)$
$+ (3 - 4.12)^2(.12) + (4 - 4.12)^2(.30)$
$+ (5 - 4.12)^2(.30) + (6 - 4.12)^2(.11)$
$+ (7 - 4.12)^2(.01) + (8 - 4.12)^2(.01)$
$= (-3.12)^2(.05) + (-2.12)^2(.10)$
$+ (-1.12)^2(.12) + (-.12)^2(.30)$
$+ (.88)^2(.30) + (1.88)^2(.11)$
$+ (2.88)^2(.01) + (3.88)^2(.01)$
$= .48672 + .44944 + .150528 + .00432$
$+ .23232 + .388784 + .082944 + .150544$
$= 1.9456$

$\sigma = \sqrt{1.9456} = 1.3948$

The variance (1.9456) is the mean squared difference between the number of sets sold and the average number of sets sold. The standard deviation (1.3948) is the typical difference between the number of sets sold and the average number of sets sold.

b) $P(\mu - \sigma \le x \le \mu + \sigma)$
$= P(4.12 - 1.3948 \le x \le 4.12 + 1.3948)$
$= P(2.7252 \le x \le 5.5148)$
$= P(x = 3, 4, \text{ or } 5) = .12 + .30 + .30 = 0.72$

c) $P(x < \mu - 2\sigma \text{ or } \mu + 2\sigma < x)$
$= P(x < 4.12 - 2(1.3948) \text{ or } 4.12 + 2(1.3948) < x)$
$= P(x < 1.3304 \text{ or } 6.9096 < x)$
$= P(x = 1, 7, \text{ or } 8) = .05 + .01 + .01 = 0.07$

6.23 $\mu = 1000(.05) + 5000(.3) + 10000(.4) + 20000(.25)$
$= 10550$
Under the royalty plan the mean amount received would be $10,550. Since this exceeds the $10,000 that would be received under the flat payment plan, the author should choose the royalty plan, if the criterion is to maximize the expected renumeration.

6.25 $\mu = 95(.05) + 96(.10) + 97(.12) + 98(.14) + 99(.24)$
$+ 100(.17) + 101(.06) + 102(.04) + 103(.03)$
$+ 104(.02) + 105(.01) + 106(.005) + 107(.005)$
$+ 108(.005) + 109(.0037) + 110(.0013) = 98.9813$

6.27 a) y is a discrete random variable since it can take on only six different values.

b) $P(y > 1.20) = .10 + .16 + .08 + .06 = .40$

$P(y < 1.40) = .36 + .24 + .10 + .16 = .86$

c) $\mu = 115.9(.36) + 118.7(.24) + 129.9(.10)$
$+ 139.9(.16) + 144.9(.08) + 159.7(.06)$
$= 41.724 + 28.488 + 12.990 + 22.384 + 11.592 + 9.582$
$= 126.760$

$$\sigma^2 = (115.9-126.760)^2(.36)$$
$$+ (118.7-126.760)^2(.24)$$
$$+ (129.9-126.760)^2(.10)$$
$$+ (139.9-126.760)^2(.16)$$
$$+ (144.9-126.760)^2(.08)$$
$$+ (159.7-126.760)^2(.06)$$
$$= 42.4583 + 15.5913 + 0.986 + 27.6255$$
$$+ 26.3248 + 65.1026 = 178.0885$$
$$\sigma = \sqrt{178.0885} = 13.345$$

The mean is the average price per gallon paid by the customers. The standard deviation is the typical difference between the price per gallon paid and the average price per gallon paid.

6.29   a) $y = 100 - 5(x)$

If $x = 1$, $y = 100 - 5 = 95$
   $x = 2$, $y = 100 - 10 = 90$
   $x = 3$, $y = 100 - 15 = 85$
   $x = 4$, $y = 100 - 20 = 80$

b)

y-value	80	85	90	95
p(y)	.1	.3	.4	.2

c) $\mu_y = 80(.1) + 85(.3) + 90(.4) + 95(.2)$
      $= 8.0 + 25.5 + 36 + 19 = 88.5$

6.31 $\mu_x^2 = 1^2(.1) + 2^2(.2) + 3^2(.3) + 4^2(.4)$
      $= .1 + .8 + 2.7 + 6.4 = 10$

6.33 n = 5, π = .25

  a) $P(x = 2) = \left(\dfrac{5!}{2!3!}\right)(.25)^2(.75)^3$
         $= 10(.0625)(.4219) = .2637$

  b) $P(x \leq 1) = P(x = 0) + P(x = 1)$
         $= \left(\dfrac{5!}{0!5!}\right)(.25)^0(.75)^5 + \left(\dfrac{5!}{1!4!}\right)(.25)(.75)^4$
         $= .2373 + .3955 = .6328$

  c) $P(x \geq 2) = 1 - P(x \leq 1) = 1 - .6328 = .3672$

  d) $P(x \neq 2) = 1 - P(x = 2) = 1 - .2637 = .7363$

6.35 n = 5, π = .5

  $P(x = 0) = \dfrac{5!}{0!5!}(.5)^0(.5)^5 = (.5)^5 = .03125$

  $P(x = 1) = \dfrac{5!}{1!4!}(.5)^1(.5)^4 = 5(.5)^5 = .15625$

  $P(x = 2) = \dfrac{5!}{2!3!}(.5)^2(.5)^3 = 10(.5)^5 = .3125$

  $P(x = 3) = \dfrac{5!}{3!2!}(.5)^3(.5)^2 = 10(.5)^5 = .3125$

  $P(x = 4) = \dfrac{5!}{4!1!}(.5)^4(.5)^1 = 5(.5)^5 = .15625$

  $P(x = 5) = \dfrac{5!}{5!0!}(.5)^5(.5)^0 = (.5)^5 = .03125$

6.37 n = 20, accept lot if x $\leq$ 1.

  a) π = .05, $P(x \leq 1)$ = .358 + .377 = .735

  b) π = .10, $P(x \leq 1)$ = .112 + .270 = .392

  c) π = .20, $P(x \leq 1)$ = .012 + .058 = .07

6.39 The expected number of trees showing damage would be 2000(.1) = 200 and the standard deviation would be $\sqrt{2000(.1)(.9)} = \sqrt{180} = 113.42$.

6.41 a) x has a binomial distribution with n = 100 and π = .2

  b) The expected score would be 100(.2) = 20.

  c) $\sigma^2 = 100(.2)(.8) = 16$; $\sigma = \sqrt{16} = 4$

  d) Most scores would be within three standard deviations of the mean (between 8 and 32). Thus, it would be highly unlikely for a person to score over 50.

6.43 a)   P(judging the coin to be biased when it is actually fair) =
           1 - P(8 $\leq$ x $\leq$ 17), where x is a binomial random variable
           with n = 25 and π = .5. From the binomial tables,
           1 - P(8 $\leq$ x $\leq$ 17)
           = 1 - (.032 + .061 + .097 + .133 + .155 + .155
             + .133 + .097 + .061 + .032)
           = 1 - .956 = .044

      b)   P(judging the coin to be fair when P(H) = .9)
           = P(8 $\leq$ x $\leq$ 17) where x is a binomial random variable with
           n = 25 and π = .9
           P(8 $\leq$ x $\leq$ 17) = .002.
           When π = P(H) = .1, P(8 $\leq$ x $\leq$17) = .002.

      c)   When π = P(H) = .6
           P(8 $\leq$ x $\leq$ 17) = .003 + .009 + .021 + .044 + .076
                              + .114 + .146 + .161 + .151 + .120
                          = .845
           When π = P(H) = .4,
           P(8 $\leq$ x $\leq$ 17) = .120 + .151 + .161 + .146 + .114
                              + .076 +.044 + .021 + .009
                              + .003
                          = .845

      d)   The probability in (a) becomes smaller and the probability
           in (b) becomes larger.

6.45 π = .9   n = 25

      a)   P(x > 20) = .138 + .227 + .226 + .199 + .072 = .902

      b)   P(x $\geq$ 20) = .065 + .138 + .227 + .266 + .199 + .072 = .967

      c) μ = 25(.9) = 22.5; σ = $\sqrt{25(.9)(.1)}$ = $\sqrt{2.25}$ = 1.5

      d)   P(x < 20) = 1 - P(x $\geq$ 20) = 1 - .967 = .033. If π = .9 then
           there is only a 3.3% chance of selecting a sample of 25
           people of which fewer than 20 favor the ban. Therefore, if
           this type of sample resulted, it would cause one to doubt
           that π $\geq$ .9.

6.47 Let x denote the number of persons who desire to purchase a Diet Coke. Then x has a binomial distribution with n = 15 and π = .6. The probability that each of the 15 is able to purchase the type of drink desired is

$$P(5 \leq x \leq 10) = P(x = 5, 6, 7, 8, 9, \text{ or } 10)$$
$$= .025 + .061 + .118 + .177 + .207 + .196$$
$$= 0.78$$

6.49 a) $\mu = 0(.10) + 1(.15) + 2(.20) + 3(.25) + 4(.20) + 5(.06)$
$\qquad + 6(.04)$
$\qquad = 0 + .15 + .40 + .75 + .80 + .30 + .24 = 2.64$

$\sigma^2 = (0-2.64)^2(.1) + (1-2.64)^2(.15)$
$\qquad + (2-2.64)^2(.20) + (3-2.64)^2(.25)$
$\qquad + (4-2.64)^2(.2) + (5-2.64)^2(.06)$
$\qquad + (6-2.64)^2(.04)$
$\qquad = .69696 + .40344 + .08192 + .0324 + .36992$
$\qquad + .334176 + .451584 = 2.3704$

$\sigma = \sqrt{2.3704} = 1.54$

b) $\mu + 3\sigma = 2.64 + 3(1.54) = 2.64 + 4.62 = 7.26$
$\mu - 3\sigma = 2.64 - 3(1.54) = 2.64 - 4.62 = -1.98$
$P(x < \mu - 3\sigma \text{ or } x > \mu + 3\sigma)$
$\qquad = P(x < -1.98 \text{ or } x > 7.26) = 0$

6.51

Numbers	1	2	3	4	w-value
Signs	+	+	+	+	+10
	+	+	+	−	+2
	+	+	−	+	+4
	+	−	+	+	+6
	−	+	+	+	+8
	+	+	−	−	−4
	+	−	+	−	−2
	+	−	−	+	0
	−	+	−	+	+2
	−	+	+	−	0
	−	−	+	+	+4
	−	−	−	+	−2
	−	−	+	−	−4
	−	+	−	−	−6
	+	−	−	−	−8
	−	−	−	−	−10

w-value	−10	−8	−6	−4	−2	0
p(w)	$\frac{1}{16}$	$\frac{1}{16}$	$\frac{1}{16}$	$\frac{2}{16}$	$\frac{2}{16}$	$\frac{2}{16}$

w-value	2	4	6	8	10
p(w)	$\frac{2}{16}$	$\frac{2}{16}$	$\frac{1}{16}$	$\frac{1}{16}$	$\frac{1}{16}$

6.53

Outcomes	x-value	probability
a	1	$\frac{1}{4}$
b	1	$\frac{1}{4}$
ca	2	$\frac{1}{12}$
cb	2	$\frac{1}{12}$
da	2	$\frac{1}{12}$
db	2	$\frac{1}{12}$
cda	3	$\frac{1}{24}$
cdb	3	$\frac{1}{24}$
dca	3	$\frac{1}{24}$
dcb	3	$\frac{1}{24}$

x-value	1	2	3
p(x)	$\frac{1}{2}$	$\frac{1}{3}$	$\frac{1}{6}$

6.55 $y = x - 4$

x-value	4	5	6	7
y-value	0	1	2	3
p(y)	.1552	.2688	.29952	.27648

6.57   a)

x	$y = profit = x(1.00) - .75$
1	.25
2	1.25
3	2.25
4	2.25
5	2.25
6	2.25

At most three can be
sold since only three
were stocked.

y	.25	1.25	2.25
p(y)	$\frac{1}{15}$	$\frac{2}{15}$	$\frac{12}{15}$

$$\mu = \left(\frac{1}{15}\right)\left(\frac{1}{4}\right) + \left(\frac{2}{15}\right)\left(\frac{5}{4}\right) + \left(\frac{12}{15}\right)\left(\frac{9}{4}\right)$$

$$= \frac{(1 + 10 + 108)}{60} = \frac{119}{60} = 1.983$$

b)

x   y = profit = x(1.00) - 1.00

x	y
1	0
2	1
3	2
4	3
5	3
6	3

At most four can be sold since only four were stocked.

y	0	1	2	3
p(y)	$\frac{1}{15}$	$\frac{2}{15}$	$\frac{3}{15}$	$\frac{9}{15}$

$$\mu = 0\left(\frac{1}{15}\right) + 1\left(\frac{2}{15}\right) + 2\left(\frac{3}{15}\right) + 3\left(\frac{9}{15}\right) = \frac{35}{15} = 2.33$$

The mean profit per week is larger if four magazines are stocked.

6.59   a) $\mu = \Sigma x\left(\frac{1}{n}\right) = \left(\frac{1}{n}\right)\Sigma x = \left(\frac{1}{n}\right)\left[\frac{n(n+1)}{2}\right] = \frac{(n+1)}{2}$

b) $\sigma^2 = \left[\Sigma x^2\left(\frac{1}{n}\right)\right] - \left[\frac{(n+1)}{2}\right]^2 = \left(\frac{1}{n}\right)\Sigma x^2 - \frac{(n+1)^2}{4}$

$$= \frac{1}{n}\frac{n(n+1)(2n+1)}{6} - \frac{(n+1)^2}{4}$$

$$= \frac{(n+1)(2n+1)}{6} - \frac{(n+1)^2}{4}$$

$$= \frac{2n^2 + 3n + 1}{6} - \frac{n^2 + 2n + 1}{4}$$

$$= \frac{8n^2 + 12n + 4 - 6n^2 - 12n - 6}{24}$$

$$= \frac{2n^2 - 2}{24} = \frac{n^2 - 1}{12}$$

c)   When n = 10, $\mu_x = \frac{(10+1)}{2} = 5.5$

$\sigma^2 = \frac{99}{12}$ and $\sigma = \sqrt{\frac{99}{12}} = 2.87$

6.61   a) $p(0) = \left(\frac{4}{5}\right)\left(\frac{3}{4}\right)\left(\frac{2}{3}\right) = \frac{2}{5} = .4$

b)   p(only Sony component selected is the receiver)

$$= \left(\frac{1}{5}\right)\left(\frac{3}{4}\right)\left(\frac{2}{3}\right) = \frac{2}{20} = .10$$

p(1) = P(only Sony component selected is a receiver,
         turntable or cassette deck)

$$= .1 + \left(\frac{4}{5}\right)\left(\frac{1}{4}\right)\left(\frac{2}{3}\right) + \left(\frac{4}{5}\right)\left(\frac{3}{4}\right)\left(\frac{1}{3}\right)$$

$$= \frac{1}{10} + \frac{2}{15} + \frac{1}{5} = \frac{13}{30} = .4333$$

c)   p(2) = P(receiver and turntable only are Sony)
        + P(receiver and cassette only are Sony)
        + P(turntable and cassette only are Sony)

$$= \left(\frac{1}{5}\right)\left(\frac{1}{4}\right)\left(\frac{2}{3}\right) + \left(\frac{1}{5}\right)\left(\frac{3}{4}\right)\left(\frac{1}{3}\right) + \left(\frac{4}{5}\right)\left(\frac{1}{4}\right)\left(\frac{1}{3}\right)$$

$$= \frac{9}{60} = .15$$

p(3) = P(receiver, turntable and cassette deck are Sony)

$$= \left(\frac{1}{5}\right)\left(\frac{1}{4}\right)\left(\frac{1}{3}\right) = \frac{1}{60}$$

x	0	1	2	3
p(x)	$\frac{24}{60}$	$\frac{26}{60}$	$\frac{9}{60}$	$\frac{1}{60}$

6.63 a) $P(3) = \dfrac{e^{-2}(2)^3}{3!} = \dfrac{.1353353(8)}{6} = .1804$

b) $P(x \geq 2) = 1-P(x < 2) = 1 - [P(0) + P(1)]$

$$= 1 - \left[\frac{e^{-2}(2)^0}{0!} + \frac{e^{-2}(2)^1}{1!}\right]$$

$$= 1 - 3e^{-2} = 1 - .406 = .594$$

c) $p(5) = \dfrac{e^{-4}(4)^5}{5!} = \dfrac{(.018315638)(1024)}{120} = .15629$

d) $\mu = 2(30) = 60$

6.65 a)   With three recruits

```
    1         1         1         1         1       1
2  3  4    2    4    2    3    2       2    3    2
    3         4         3    4             4    3
                                               4
```

b)   Let x = number of leaves on a tree. x can take on the values
1, 2, or 3.

c)

x	1	2	3
p(x)	$\frac{1}{6}$	$\frac{4}{6}$	$\frac{1}{6}$

$$\mu = \Sigma(x) = 1\left(\frac{1}{6}\right) + 2\left(\frac{4}{6}\right) + 3\left(\frac{1}{6}\right) = \frac{12}{6} = 2$$

d)   number of trees = (number of recruits)!
So, for four recruits, the number of trees would be 4! = 24.

Section 7.1

7.1 a)   P(at most 5 minutes elapses before dismissal)
          = (1/10)(5-0) = .5

     b)   P(3 ≤ x ≤ 5) = (1/10)(5-3) = .2

     c) μ = 5. Because the density function is uniform on the
          interval 0 to 10, the mean should be at the middle of the
          interval.

7.3 a)   density

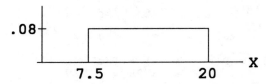

     b)   height = $\dfrac{1}{(20-7.5)}$ = $\dfrac{1}{12.5}$ = .08

     c)   P(x ≤ 12) = .08(12-7.5) = .08(4.5) = .36

     d)   P(10 ≤ x ≤ 15) = .08(15-10) = .08(5) = .40
          P(12 ≤ x ≤ 17) = .08(17-12) = .08(5) = .40
          They are equal because the lengths of the intervals
          describing the events of interest are equal.

     e) μ = $\dfrac{(20+7.5)}{2}$ = 13.75

     f) σ = $\dfrac{(20-7.5)}{\sqrt{12}}$ = $\dfrac{12.5}{\sqrt{12}}$ = 3.608

7.5 a)   P(x ≤ 10) = (10 - 0)(.05) = .5
          P(x ≥ 15) = (20 - 15)(.05) = .25

     b)   P(7 ≤ x ≤ 12) = (12 - 7)(.05) = .25

     c)   P(x ≤ c) = (c)(.05) = .9 which leads to c = .9/.05 = 18

7.7 a)   Area to the left of 1 = .5 + .75(1 - 1/3) = 1/2 + 1/2 = 1

     b)   P(x ≤ .5) = .5 + .75(.5 - (.5)³/3)
                    = 1/2 + (3/4)[1/2 - 1/24] = 1/2 + (3/4)[11/24]
                    = 1/2 + 33/96 = 81/96 = .84375

     c)   P(-.5 < x < .5) = P(x < 5) - P(x < -.5)
                          = .84375 - [1/2 + (3/4)(-1/2 + 1/24)]
                          = .84375 - [1/2 + (3/4)(-11/24)]
                          = .84375 - 15/96
                          = .84375 - .15625 = .6875

     d) μ = 0 (the center of the interval from -1 to 1)

7.9   a) $P(z < 1.75) = .9599$

      b) $P(z < -.68) = .2483$

      c) $P(z > 1.20) = 1 - P(z < 1.20) = 1 - .8849 = .1151$

      d) $P(z > -2.82) = 1 - P(z < -2.82) = 1 - .0024 = .9976$

      e) $P(-2.22 < z < .53) = P(z < .53) - P(z < -2.22)$
           $= .7019 - .0132 = .6887$

      f) $P(-1 < z < 1) = P(z < 1) - P(z < -1) = .8413 - .1587 = .6826$

      g) $P(-4 < z < 4) = P(z < 4) - P(z < -4) = 1 - 0 = 1$

7.11 a)   .9909   b)   .9909   c)   .1093

      d) $P(1.14 < z < 3.35) = P(z < 3.35) - P(z < 1.14)$
           $= .9996 - .8729 = .1267$

      e) $P(-.77 \leq z \leq -.55) = P(z < -.55) - P(z < -.77)$
           $= .2912 - .2206 = .0706$

      f) $P(-2.90 < z \leq 1.15) = P(z < 1.15) - P(z < -2.90)$
           $= .8749 - .0019 = .8730$

      g) $P(2 < z) = 1 - P(z < 2) = 1 - .9772 = .0228$

      h) $P(-3.38 < z) = 1 - P(z < -3.38) = 1 - .0004 = .9996$

      i) $P(z < 4.98) = 1$

7.13 a) $c = .23$ b) $c = -.23$ c) $c = 2.75$

      d) $c = 1.16$

7.15 a) $c = 1.96$ b) $c = 1.28$ c) $c = -1.28$

      d) $c = 2.58$ e) $c = 2.58$ f) $c = 3.10$

7.17 a) $P(x < 5.0) = P(z < (5-5)/.2) = P(z < 0) = .5$

      b) $P(x < 5.4) = P(z < (5.4-5)/.2) = P(z < 2) = .9772$

      c) $P(x \leq 5.4) = P(z \leq (5.4-5)/.2) = P(z \leq 2) = .9772$

      d) $P(4.6 < x < 5.2) = P((4.6-5)/.2 < z < (5.2-5)/.2)$
           $= P(-2 < z < 1) = P(z < 1) - P(z < -2)$
           $= .8413 - .0228 = .8185$

      e) $P(4.5 < x) = P((4.5-5)/.2 < z) = P(-2.5 < z)$
           $= 1 - P(z < -2.5) = 1 - .0062 = .9938$

      f) $P(4.0 < x) = P((4-5)/.2 < z) = P(-5 < z)$
           $= 1 - P(z < -5) = 1 - 0 = 1$

7.19 a) $P(x < 14.8) = P(z < (14.8-15)/.1) = P(z < -2) = .0228$

92

b) $P(14.7 < x < 15.1) = P((14.7-15)/.1 < z < (15.1-15)/.1)$
$= P(-3 < z < 1) = .8413 - .0013 = .8400$

7.21 a) $P(x \leq 17) = P(z < (17-15)/1.25) = P(z < 1.6) = .9452$

b) $P(x < 17) = P(z < 1.6) = .9452$

c) $P(12 < x < 17) = P((12-15)/1.25 < z < (17-15)/1.25)$
$= P(-2.4 < z < 1.6) = .9452 - .0082 = .9370$

d) $P(x < 15 - 2(1.25)$ or $x > 15 + 2(1.25))$
$= P(z < -2$ or $z > 2) = P(z < -2) + [1 - P(z < 2)]$
$= .0228 + 1 - .9772 = .0456$

7.23 a) $c = 30 - 2.33(.6) = 30 - 1.398 = 28.602$

b) $[(30 + c) - 30]/.6 = 1.96 \Rightarrow c = 1.96(.6) = 1.176$

7.25 a) $P(x > 30.5) = P(z > (30.5-31/.2)) = P(z > -2.5)$
$= 1 - P(z < -2.5) = 1 - .0062 = .9938$

b) $P(30.5 < x < 31.5) = P((30.5-31)/.2 < z < (31.5-31)/.2)$
$= P(-2.5 < z < 2.5) = .9938 - .0062 = .9876$

$P(30 < x < 32) = P((30-31)/.2 < z < (32-31)/.2)$
$= P(-5 < z < 5) = 1 - 0 = 1$

c) $P(\text{one tire being underinflated}) = P(x < 30.4)$
$= P(z < (30.4-31)/.2) = P(z < -3) = .0013$

$P(\text{at least one of the four tires is underinflated})$
$= 1 - P(\text{none are underinflated})$
$= 1 - P(.9987)^4 = 1 - .9948 = .0052$

7.27 $P(\text{not acceptable}) = 1 - P(.496 < x < .504)$
$= 1 - P((.496-.499)/.002 < z < (.504-.499)/.002)$
$= 1 - P(-1.5 < z < 2.5) = 1 - (.9938 - .0668) = .073$

7.29 a) $P(x \leq \mu + 2.33\sigma) = P(z < ((\mu + 2.33\sigma - \mu)/\sigma)$
$= P(z < 2.33) = .9901$

b)   95th percentile $= 31 + 2.33(.2) = 31.466$

c)   90th percentile $= 31 + 1.28(.2) = 31.256$

d)   1st percentile $= 31 - 2.33(.2) = 30.534$

7.31 a) $P(x \leq 120) \approx P(z \leq (120.5-150)/10) = P(z < -2.95) = .0016$

      b) $P(125 \leq x) \approx P((124.5-150)/10 < z) = P(-2.55 < z)$
             $= 1 - P(z < -2.55) = 1 - .0054 = .9946$

      c) $P(135 \leq x \leq 160) \approx P((134.5-150)/10 < z < (160.5-150)/10)$
             $= P(-1.55 < z < 1.05) = .8531 - .0606 = .7925$

7.33 a) $P(x = 30) \approx P((29.5-30)/3.4641 < z < (30.5-30)/3.4641)$
             $P(-.14 < z < .14) = .5557 - .4443 = .1114$

      b) $P(x = 25) \approx P((24.5-30)/3.4641 < z < (25.5-30)/3.4641)$
             $= P(-1.59 < z < -1.30) = .0968 - .0559 = .0409$

      c) $P(x \leq 25) \approx P(z < (25.5-30)/3.4641) = P(z < -1.30) = .0968$

      d) $P(25 \leq x \leq 40) \approx P((24.5-30)/3.4641 < z < (40.5-30)/3.4641)$
             $= P(-1.59 < z < 3.03) = .9988 - .0559 - .9429$

      e) $P(25 < x < 40)$
             $\approx P((25.5 - 30)/3.4641 < z < (39.5-30)/3.4641)$
             $= P(-1.30 < z < 2.74) = .9969 - .0968 = .9001$

7.35 $\mu = 100(.25) = 25$ $\sigma = \sqrt{100(.25)(.75)} = 4.3301$

      a) $P(20 \leq x \leq 30) \approx P((19.5-25)/4.3301 < z < (30.5-25)/4.3301)$
             $= P(-1.27 < z < 1.27) = .8980 - .1020 = .7960$

      b) $P(20 < x < 30) \approx P((20.5-25)/4.3301 < z < (29.5-25)/4.3301)$
             $= P(-1.04 < z < 1.04) = .8508 - .1492 = .7016$

      c) $P(35 \leq x) \approx P(34.5-25)/4.3301 < z) = P(2.19 < z)$
             $= 1 - P(z < 2.19) = 1 - .9857 = .0143$

      d) $\mu - 2\sigma = 25 - 2(4.3301) = 16.3398$
             $\mu + 2\sigma = 25 + 2(4.3301) = 33.6602$
             $P(x < 16.3398 \text{ or } x > 33.6602) = P(x \leq 16 \text{ or } x \geq 34)$
             $= 1 - P(17 \leq x \leq 33)$
             $\approx 1 - P((16.5-25)/4.3301 < z < (33.5-25)/4.3301)$
             $= 1 - P(-1.96 < z < 1.96) = 1 - (.975 - .025) = .05$

7.37 a) $n\pi = 50(.05) = 2.5$ which does not exceed 5. Hence, the
             normal approximation to the binomial should not be used.

      b) $n\pi = 500(.05) = 25 \geq 5$; $n(1-\pi) = 475 \geq 5$. Hence the normal
             approximation to the binomial could be used.
             $\mu = 25$ $\sigma = \sqrt{500(.05)(.95)} = 4.8734$
             $P(20 \leq x) \approx P((19.5-25)/4.8734 < z) = P(-1.13 < z)$
             $= 1 - P(z < -1.13) = 1 - .1292 = .8708$

7.39 Since this plot appears to be very much like a straight line, it
      is reasonable to conclude that the normal distribution provides an
      adequate description of the steam rate distribution.

7.41

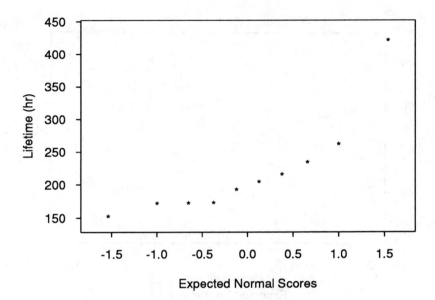

Since the graph exhibits a pattern substantially different from
that of a straight line, one would conclude that the distribution
of the variable "component lifetime" cannot be adequately modeled
by a normal distribution.

7.43

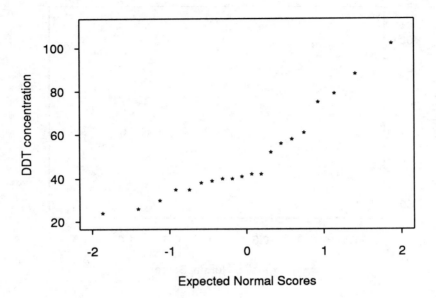

Since the graph exhibits a pattern similar to that of a straight line (at least there is no glaring difference), one would conclude that the distribution could be adequately modeled by a normal distribution.

7.45 a) $P(.525 < x < .550) = P((.525-.5)/.025 < z < (.550-.5)/.025)$
$= P(1 < z < 2) = .9772 - .8413 = .1359$

b) $P(\mu + 2\sigma < x) = P(2 < z) = 1 - P(x < 2) = 1 - .9772 = .0228$

c) $P(.475 < x\ ) = P((.475-.5)/.025 < z) = P(-1 < z)$
$= 1 - P(z < -1) = 1 - .1587 = .8413$
P(all three have at least .475 ounces of cheese)
$= (.8413)^3 = .5955$

7.47 a) $P(45 < x) = P((45-60)/10 < z) = P(-1.5 < z)$
$= 1 - P(z < -1.5) = 1 - .0668 = .9332$

b) $c = \mu + 1.28\sigma = 60 + 1.28(10) = 60 + 12.8 = 72.8$

c) Expected revenue = 10 + 50 = $60

7.49 $P(x < 4.9) = P(z < (4.9-5)/.05) = P(z < -2) = .0228$
$P(5.2 < x) = P((5.2-5)/.05 < z) = P(4 < z) \approx 0$

7.51 a) $P(x < 5'7") = P(x < 67") = P(z < (67-66)/2) = P(z < .5)$
$= .6915$
No, the claim that 94% of all women are shorter than 5'7" is
not correct. Only about 69% of all women are shorter than
5'7".

b) About 69% of adult women would be excluded from employment
due the height requirement.

7.53 a) $P(5.9 < x < 6.15) = P((5.9-6)/.1 < z < (6.15-6)/.1)$
$= P(-1 < z < 1.5) = .9332 - .1587 = .7745$

b) $P(6.1 < x) = P((6.1-6)/.1 < z) = P(1 < z) = 1 - P(z < 1)$
$= 1 - .8413 = .1587$

c) $P(x < 5.95) = P(z < (5.95-6)/.1) = P(z < -.5) = .3085$

d) 95th percentile = 6 + 1.645(.1) = 6.1645

7.55 a) $\mu = 200(.16) = 32;\ \sigma = \sqrt{200(.16)(.84)} = \sqrt{26.88} = 5.1846$

b) $P(25 \le x \le 40) \approx P((24.5-32)/5.1846 < z < (40.5-32)/5.1846)$
$P(-1.45 < z < 1.64) = .9495 - .0735 = .8760$

c) $P(50 < x) \approx P((50.5-32)/5.1846 < z) = P(3.57 < z)$
$= 1 - P(3.57) = 1 - 1 = 0$
Yes, the 16% figure would be doubted because the probability
is nearly zero that more than 50 would be found to be
uninsured, if the 16% figure were correct.

7.57 P(a cork meets specifications) = $P(2.9 < x < 3.1)$
$= P((2.9-3)/.1 < z < (3.1-3)/.1) = P(-1 < z < 1)$
$= .8413 - .1587 = .6826$
Therefore, the proportion of corks produced by this machine that
will be defective is 1 - .6826 - .3174.

7.59 a) $P(x < 620) = P(z < (620-500)/100) = P(z < 1.2) = .8849$
Thus, 620 is about the 88th percentile.

b) $P(x < 710) = P(z < (710-500)/100) = P(z < 2.1) = .9821$
     Thus, 710 is about the 98th percentile.

c)   90th percentile = $500 + 1.28(100) = 628$
     To be at the 90th percentile, a student would have to have a
     score of 628.

7.61 $P(x < w - 1) = .99 \Rightarrow (w - 1 - 10)/2 = 2.33 \Rightarrow$
     $w = 11 + 2(2.33) = 15.66$

7.63 a) $P(x \leq 2000) = 1 - e^{-2000/1000} = 1 - e^{-2} = 1 - .1353 = .8647$

     b) $P(x \leq 1000) = 1 - e^{-1000/1000} = 1 - e^{-1} = 1 - .3679 = .6321$
          This probability is not .5, because the density curve is not
          symmetric.

     c) $P(500 \leq x \leq 2000) = P(x \leq 2000) - P(x \leq 500)$
          $= .8647 - [1 - e^{-500/1000}] = .8647 - [1 - .6065]$
          $= .8647 - .3935 = .4712$

CHAPTER 8
SAMPLING DISTRIBUTIONS

Section 8.1

8.1   A statistic is any quantity computed from the observations in a sample. A population characteristic is a quantity which describes the popoulation from which the sample was taken.

8.3   a)   population characteristic   b)   statistic
        c)   population characteristic   d)   statistic
        e)   statistic

8.5   a)   First, obtain a list of all doctors practicing in Los Angeles County. Next, write the name of each doctor on a different card. Then shuffle the cards well, and finally, select the desired number of cards (doctors) to compose the sample.

        b)   First, obtain a list of all students enrolled in the university. Next, write the name of each student on a different card. Shuffle the cards well, and finally, select the desired number of cards (students) for inclusion in the sample.

        c)   First, obtain a listing of all possible box locations in the warehouse. Write each location on a different card. Shuffle the cards well, and then select the desired number of cards. The boxes located at those locations specified on the selected cards make up the sample.

        d)   Obtain a list of all registered voters in the community. Write the name of each voter on a different card. Shuffle the cards well, and select the desired number of cards (voters).

        e)   Obtain a list of all subscribers to the newspaper. Write the name of each subscriber on a different card. Shuffle the cards well, and select the desired number of cards (subscribers).

        f)   Number the radios 1 through 1000. Write the numbers 1 through 1000 on different cards, and shuffle the cards well. Select the desired number of cards (radios) for inclusion in the sample.

8.7   No unique solution.

8.9   a)

Sample	Value of t	Sample	Value of t
1,5	6	5,10	15
1,10	11	5,20	25
1,20	21	10,20	30

t value	6	11	15	21	25	30
probability	1/6	1/6	1/6	1/6	1/6	1/6

b) The population mean is $\mu = (1+5+10+20)/4 = 36/4 = 9$

$\mu_t = (1/6)(6) + (1/6)(11) + (1/6)(15) + (1/6)(21)$
$+ (1/6)(25) + (1/6)(30) = 108/6 = 18$

μ_t is twice the value of μ.

8.11

Sample	Value of $\overline{x}$	Value of Median	Value of statistic #3
2,3,3	2.67	3	2.5
2,3,4	3	3	3
3,3,4	3.33	3	3.5
3,4,4	3.67	4	3.5
2,3,4	3	3	3
2,3,4	3	3	3
3,3,4	3.33	3	3.5
2,3,4	3	3	3
2,4,4	3.33	4	3
3,4,4	3.67	4	3.5

Sampling distribution of Statistic #1.

value of $\overline{x}$	2.67	3	3.33	3.67
probability	.1	.4	.3	.2

Sampling distribution of statistic #2.

value	3	4
probability	.7	.3

Sampling distribution of statistic #3.

value	2.5	3	3.5
probability	.1	.5	.4

(Each student's discussion will differ).

8.13   a)

Sample	$\overline{x}$ value	Sample	$\overline{x}$ value
212,379	295.5	379,350	364.5
212,350	281	379,575	477
212,575	393.5	350,575	462.5

$\overline{x}$ value	281	295.5	364.5	393.5	462.5	477
probability	1/6	1/6	1/6	1/6	1/6	1/6

$\mu_{\overline{x}}$ = (1/6)(281) + (1/6)(295.5) + (1/6)(364.5) + (1/6)(393.5)
        + (1/6)(462.5) + (1/6)(477)
    = 46.833 + 49.25 + 60.75 + 65.583 + 77.083 + 79.5 = 379

b)

Sample	$\overline{x}$ value
212,350	281
212,575	393.5
379,350	364.5
379,575	477

$\overline{x}$ value	281	364.5	393.5	477
probability	.25	.25	.25	.25

$\mu_{\overline{x}}$ = (1/4)(281) + (1/4)(364.5) + (1/4)(393.5) + (1/4)(477)
    = 70.25 + 91.125 + 98.375 + 119.25 = 379

8.15 a) $\sigma^2 = \frac{1}{5}[(8-11.8)^2+(14-11.8)^2+(16-11.8)^2 + (10-11.8)^2$

$+ (11-11.8)^2]$

$= \frac{1}{5}[(-3.8)^2+(2.2)^2+(4.2)^2+(-1.8)^2+(-.8)^2]$

$= \frac{1}{5}[14.44 + 4.84 + 17.64 + 3.24 + 0.64]$

$= \frac{1}{5}(40.8) = 8.16$

b) One random sample of size 2 consists of the elements 14 and 16. s^2 for this sample would be

$(14 - 15)^2 + (16 - 15)^2 = (-1)^2 + (1)^2 = 2$

c) Twenty four additional sample of size 2 and their sample variances are:

Sample	Value of s^2	Sample	Value of s^2
8,11	4.5	16,14	2.0
14,11	4.5	11,8	4.5
16,10	18.0	8,10	2.0
10,14	8.0	8,16	32.0
16,14	2.0	10,14	8.0
10,11	0.5	10,14	8.0
10,8	2.0	16,14	2.0
11,14	4.5	10,8	2.0
8,14	18.0	11,10	0.5
10,14	8.0	8,11	4.5
10,8	2.0	14,16	2.0
16,14	2.0	14,16	2.0

d)

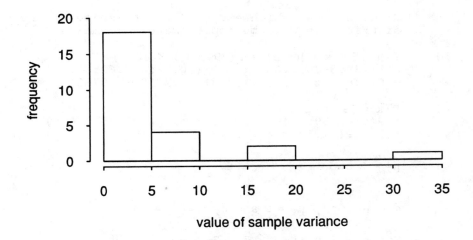

For this group of 25 samples of size two, the sample variances are either close to the population variance or smaller than the population variance. There are only three sample variances that are larger than the population variance. The sample variances do differ quite a bit in value from one sample to the next.

8.17   If samples of size ten rather than size five had been used, the histograms would be similar in that they both would be centered close to 260.25. They would differ in that the histogram based on n = 10 would have less variability than the histogram based on n = 5.

8.19   For n = 36, 50, 100, and 400

8.21   a) $\mu_{\overline{x}} = 40$, $\sigma_{\overline{x}} = \dfrac{5}{\sqrt{64}} = \dfrac{5}{8} = .625$

Since n = 64, which exceeds 30, the shape of the sampling distribution will be approximately normal.

b) $P(\mu-.5 < \overline{x} < \mu+.5) = P(39.5 < \overline{x} < 40.5)$
= P((39.5-40)/.625 < z < (40.5-40)/.625)
= P(-.8 < z < .8) = .7881 - .2119 = .5762

c) $P(\overline{x} < 39.3 \text{ or } \overline{x} > 40.7)$
= P(z < (39.3-40)/.625 or z > (40.7-40)/.625)
= 1 - P(-1.12 < z < 1.12) = 1 - [.8686 - .1314] = .2628

8.23   a) $\mu_{\overline{x}} = 2$ and $\sigma_{\overline{x}} = \dfrac{.8}{\sqrt{9}} = .267$

b)   For n = 20, $\mu_{\overline{x}} = 2$ and $\sigma_{\overline{x}} = \dfrac{.8}{\sqrt{20}} = .179$

For n = 100, $\mu_{\overline{x}} = 2$ and $\sigma_{\overline{x}} = \dfrac{.8}{\sqrt{100}} = .08$

In all three cases $\mu_{\overline{x}}$ has the same value, but the standard deviation of $\overline{x}$ decreases as n increases. A sample of size 100 would be most likely to result in an $\overline{x}$ value close to μ. This is because the sampling distribution of $\overline{x}$ for n = 100 has less variability than those for n = 9 or 20.

8.25   a) $\sigma_{\overline{x}} = \dfrac{10}{\sqrt{100}} = 1$
$\overline{x}$ will be within 2 of the value of μ if $\mu-2 \le \overline{x} \le \mu+2$
$P(\mu - 2 \le \overline{x} \le \mu + 2)$
= P[((\mu -2) - \mu)/1 < z < ((\mu + 2) - \mu)/1)]
= P(-2 < z < 2) = .9772 - .0228 = .9544

b)   Approximately 95% of the time, $\overline{x}$ will be within $2\sigma_{\overline{x}} = 2(1) = 2$ of μ.
Approximately .3% of the time, $\overline{x}$ will be further than $3\sigma_{\overline{x}} = 3(1) = 3$ from μ.

8.27 a)

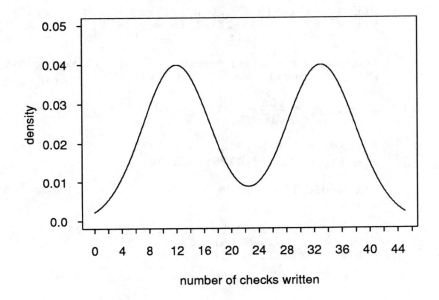

b) $\mu_{\overline{x}} = 22.0$, $\sigma_{\overline{x}} = \dfrac{16.5}{\sqrt{100}} = 1.65$

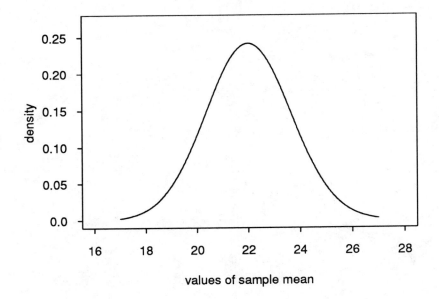

c)   The central limit theorem is the justification for the
     following computations.

$P(\overline{x} \leq 20) = P(z \leq (20-22)/1.65) = P(z \leq -1.21) = .1131$
$P(\overline{x} \geq 25) = P((25-22)/1.65 < z) = P(1.82 < z)$
$\qquad\qquad = 1 - P(z \leq 1.82) = 1 - .9656 = .0344$

8.29 a)   Because the x distribution is normal, the $\overline{x}$ distribution
           will be normal (even for n less than 30).

$\mu_{\overline{x}} = 3; \; \sigma_{\overline{x}} = \dfrac{.05}{\sqrt{16}} = .0125$

b) $3 + 3\sigma_{\overline{x}} = 3 + 3(.0125) = 3.0375$
     $3 - 3\sigma_{\overline{x}} = 3 - 3(.0125) = 2.9625$

c) $P(\overline{x} \text{ would be outside } 3 \pm 3\sigma_{\overline{x}}) = 1 - (.9987 - .0013)$
     $= .0026$

d) $P(\text{the problem is detected}) = P(\overline{x} < 2.9625 \text{ or } \overline{x} > 3.0375)$
     $= P(z < (2.9625-3.05)/.0125 \text{ or } z > (3.0375-3.05)/.0125)$
     $= P(z < -7 \text{ or } z > -1) = P(z < -7) + P(z > -1)$
     $= 0 + (1 - .1587) = .8413$

8.31 When π = .65
      n = 10, nπ = 10(.65) = 6.5 $\geq$ 5, n(1-π) = 10(.35) = 3.5 < 5
      n = 20, nπ = 20(.65) = 13 $\geq$ 5, n(1-π) = 20(.35) = 7 $\geq$ 5
      So n = 20, 30, 50, 100, and 200.

      When π = .2
      n = 10, nπ = 10(.2) = 2 < 5, n(1-π) = 10(.8) = 8 $\geq$ 5
      n = 20, nπ = 20(.2) = 4 < 5, n(1-π) = 20(.8) = 16 $\geq$ 5
      n = 30, nπ = 30(.2) = 6 $\geq$ 5, n(1-π) = 30(.8) = 24 $\geq$ 5
      So n = 30, 50, 100, and 200.

8.33 a)   If n = 10, then nπ = 10(.55) = 5.5 $\geq$ 5 and n(1 - π) =
            10(.45) = 4.5. The latter is not 5 or more, and therefore
            the sampling distribution of p, based on samples of size 10,
            would not be well described by a normal curve.

      b)   For n = 400, μ_p = .55 and σ_p = $\sqrt{\dfrac{.55(.45)}{400}}$ = .0249

      c)   P(.5 $\leq$ p $\leq$ .6) = P((.5-.55)/.0249 < z < (.6 - .55)/.0249)
            = P(-2.01 < z < 2.01) = .9778 - .0222 = .9556

      d) μ_p = .4, σ_p = $\sqrt{\dfrac{.4(.6)}{400}}$ = .0245,
            P(.5 < p) = P((.5-.4)/.0245 < z) = P(4.1 < z)
            = 1 - P(z $\leq$ 4.1) = 1 - 1 = 0

8.35 μ_p = π = .48, σ_p = $\sqrt{\dfrac{(.48)(.52)}{500}}$ = .02234
      P(.5 < p) = P((.5-.48)/.02234 < z) = P(.90 < z)
      = 1 - P(z < .9) = 1 - .8159 = .1841

8.37 $\mu_{\overline{x}} = .8$ and $\sigma_{\overline{x}} = \dfrac{.1}{\sqrt{100}} = .01$

$P(\overline{x} < .79) = P(z < (.79-.8)/.01) = P(z < -1) = .1587$
$P(\overline{x} < .77) = P(z < (.77-.8)/.01) = P(z < -3) = .0013$

8.39 $\pi = .20$, $n = 100$, $\mu_p = .2$, $\sigma_p = \sqrt{\dfrac{(.2)(.8)}{100}} = .04$

$P(.25 < p) = P((.25-.20)/.04 < z) = P(1.25 < z)$
$= 1 - P(z \leq 1.25) = 1 - .8944 = .1056$

8.41 a) $P(850 < x < 1300)$
$= P((850-1000)/150 < z < (1300-1000)/150)$
$= P(-1 < z < 2) = .9772 - .1587 = .8185$

b) $\mu_{\overline{x}} = 1000$, $\sigma_{\overline{x}} = \dfrac{150}{\sqrt{10}} = 47.43$

$P(950 < \overline{x} < 1100)$
$= P((950-1000)/47.53 < z < (1100-1000)/47.43))$
$= P(-1.05 < z < 2.11) = .9826 - .1469 = .8357$

$P(850 < \overline{x} < 1300)$
$= P((850-1000/47.43 < z < (1300-1000)/47.43)$
$= P(-3.2 < z < 6.3) = 1 - .0007 = .9993$

8.43 $\mu = 100$, $\sigma = 30$, $\mu_{\overline{x}} = 100$, $\sigma_{\overline{x}} = \dfrac{30}{\sqrt{50}} = 4.2426$

$P(5300 < \text{total}) = P(106 < \overline{x}) = P((106-100)/4.2426 < z)$
$= P(1.41 < z) = 1 - P(z \leq 1.41) = 1 - .9207 = .0793$

9.1   Statistic II would be preferred because it is unbiased and has smaller variance than the other two.

9.3   A point estimate of the proportion of all U.S. doctors who oppose the sale of organs is $\frac{184}{244} = 0.7541$

9.5   a) $\overline{x} = 4.84/10 = .484$

       b)   p = (# exceeding .5)/10 = 3/10 = .3

9.7   The point estimate of π would be

       $$p = \frac{\text{number in sample registered}}{n} = \frac{14}{20} = .70$$

9.9   An estimate of μ_d is $\overline{x}_J - \overline{x}_A$. From 10a) $\overline{x}_J = 120.6$. From this data, $\overline{x}_A = 47.4$. Thus, an estimate of μ_d is (120.6-47.4) = 73.2

9.11 a)   As the confidence level increases, the width of the large sample confidence interval also increases.

      b)   As the sample size increases, the width of the large sample confidence interval decreases.

      c)   As the population standard deviation increases, the width of the large sample confidence interval also increases.

9.13 a)   The 90% confidence interval would have been narrower, since its z critical value would have been smaller.

      b)   The statement is incorrect. The 95% refers to the percentage of <u>all possible</u> samples that result in an interval which includes μ, not to the chance (probability) that a specific interval contains μ.

      c)   Again this statement is incorrect. While we would expect <u>close</u> to 95 of the 100 intervals constructed to contain μ, we cannot be <u>certain</u> that exactly 95 out of 100 of them will. The 95% refers to the percentage of <u>all possible</u> intervals that include μ.

9.15 The 99% confidence interval would be

$$8.20 \pm (2.58)\left(\frac{4.5}{\sqrt{62}}\right) \rightarrow 8.20 \pm 1.47 \rightarrow (6.73, 9.67)$$

That is, based on the data is this sample, plausible values for the average optimum amount of allocated time per hour for commercials during prime time are those between 6.73 minutes and 9.67 minutes.

9.17 n = 64, $\overline{x}$ = 352 minutes, s = 8 minutes

$$\overline{x} \pm (z \text{ critical})\frac{s}{\sqrt{n}} = 352 \pm (2.58)\frac{8}{\sqrt{64}}$$
$$= 352 \pm 2.58 = (349.42, 354.58)$$

Therefore, with confidence of 99% the true average playing time for "6-hour" tapes is estimated to be between 349.42 minutes and 354.58 minutes. Dividing the endpoints of the interval by 60 converts it from minutes to hours, and the interval (5.8237, 5.9097) is obtained. The interval constructed extends from 5.8237 hours to 5.9097 hours, and this interval is the interval of plausible values for the true average playing time of "6-hour" tapes made by this manufacturer. Since the interval does not contain 6, the evidence points to the conclusion that the true average playing time is a value less than 6. Hence, the manufacturer could be accused of false advertising.

9.19 An estimate σ is (700-50)/4 = 650/4 = 162.5. The required sample size is

$$n = \left[\frac{1.96(162.5)}{10}\right]^2 = (31.85)^2 = 1014.42 \text{ or } n = 1015$$

9.21 The required sample size would be

$$n = \left[\frac{1.96(2000)}{500}\right]^2 = (7.84)^2 = 61.46 \text{ or } n = 62$$

9.23 $\overline{x} = 74$, $s = 40$, $n = 144$

$$\overline{x} \pm (z \text{ critical})\frac{s}{\sqrt{n}} = 74 \pm (1.645)\frac{40}{\sqrt{144}}$$

$$= 74 \pm 5.48 = (68.52, 79.48)$$

9.25 $n = 53$ boys, $m = 46$ girls, total = 99 children

a) The interval for boys is:
$$\overline{x} \pm (z \text{ critical})\frac{s}{\sqrt{n}} = 101.7 \pm (1.645)\frac{9.8}{\sqrt{53}}$$
$$= 101.7 \pm 2.214 = (99.486, 103.914).$$

The interval for girls is:
$$\overline{x} \pm (z \text{ critical})\frac{s}{\sqrt{m}} = 101.8 \pm (1.645)\frac{9.8}{\sqrt{46}}$$
$$= 101.8 \pm 2.377 = (99.423, 104.177).$$

The intervals are very similar. That is, they yield essentially the same range of plausible values. Since they do, the evidence in the samples points to the conclusion that there is no difference between the true average systolic blood pressure for boys, and the true average systolic blood pressure for girls.

b) $$\overline{x} \pm (z \text{ critical})\frac{s}{\sqrt{n}} = 86.7 \pm 2.58\left(\frac{10.3}{\sqrt{53}}\right)$$
$$= 86.7 \pm 3.65 = (83.05, 90.35)$$

9.27 $$n = \left[\frac{z \text{ critical})\sigma}{B}\right]^2 = \left[\frac{1.96(1)}{.1}\right]^2$$

$$= (19.6)^2 = 384.16. \text{ Hence, } n \text{ should be } 385.$$

9.29 For the interval to be appropriate, $np \geq 5$, $n(1-p) \geq 5$ must be satisfied.

    a) $np = 50(.3) = 15 \geq 5$, $n(1-p) = 50(.7) = 35 \geq 5$, yes

    b) $np = 50(.05) = 2.5 < 5$, no

    c) $np = 15(.45) = 6.75 \geq 5$, $n(1-p) = 15(.55) = 8.25 \geq 5$, yes

    d) $np = 100(.01) = 1 < 5$, no

    e) $np = 100(.70) = 70 \geq 5$, $n(1-p) = 100(.3) = 30 \geq 5$, yes

    f) $np = 40(.25) = 10 \geq 5$, $n(1-p) = 40(.75) = 30 \geq 5$, yes

    g) $np = 60(.25) = 15 \geq 5$, $n(1-p) = 60(.75) = 45 \geq 5$, yes

    h) $np = 80(.10) = 8 \geq 5$, $n(1-p) = 80(.9) = 72 \geq 5$, yes

9.31 The 95% confidence interval would be

$$.77 \pm 1.96\sqrt{\frac{.77(.23)}{800}} \rightarrow .77 \pm 0.29 \rightarrow (.741, .799)$$

9.33 $p = 116/162 = .716$
The 99% confidence interval that results from this data is

$$.716 \pm 2.58\sqrt{\frac{.716(.284)}{162}} \rightarrow .716 \pm .091 \rightarrow (.625, .807)$$

9.35 $p = 528/3063 = .1724$, $np = 528 \geq 5$, $n(1-p) = 2535 \geq 5$
The 90% confidence interval is:

$$p \pm (z \text{ critical})\sqrt{\frac{p(1-p)}{n}} = .1724 \pm (1.645)\sqrt{\frac{.1724(.8276)}{3063}}$$

$$= .1724 \pm .0112 = (.1612, .1836)$$

9.37 a) $n = 200$, $p = 110/200 = .55$

$$p \pm (z \text{ critical})\sqrt{\frac{p(1-p)}{n}} = .55 \pm 1.645\sqrt{\frac{.55(.45)}{200}}$$
$$= .55 \pm .058 = (.492, .608)$$

    b) $n = 200$, $p = 96/200 = .48$

$$p \pm (z \text{ critical})\sqrt{\frac{p(1-p)}{n}} = .48 \pm 1.96\sqrt{\frac{.48(.52)}{200}}$$
$$= .48 \pm .069 = (.411, .549)$$

9.39 $p = 35/63 = .5556$, $np = 35 \geq 5$, $n(1-p) = 28 \geq 5$
The 99% confidence interval is:

$$p \pm (z \text{ critical})\sqrt{\frac{p(1-p)}{n}} = .5556 \pm 2.58\sqrt{\frac{.5556(.4444)}{63}}$$

$$= .5556 \pm .1615 = (.3941, .7171)$$

9.41 $p = 43/100 = .43$, $np = 43 \geq 5$, $n(1-p) = 57 \geq 5$
The 90% confidence interval is given by:

$$p \pm (z \text{ critical})\sqrt{\frac{p(1-p)}{n}} = .43 \pm 1.645\sqrt{\frac{.43(.57)}{100}}$$
$$= .43 \pm .08 = (.35, .51)$$

9.43 a) $p = .48$, $np = 200(.48) = 96 \geq 5$, $n(1-p) = 104 \geq 5$
The 95% confidence interval is given by:

$$p \pm (z \text{ critical})\sqrt{\frac{p(1-p)}{n}} = .48 \pm 1.96\sqrt{\frac{.48(.52)}{200}}$$
$$= .48 \pm .069 = (.411, .549)$$

b) No, because it is the mean μ that is to be estimated, rather than a proportion π. The correct formula to use would be

$$\overline{x} \pm (z \text{ critical})\frac{s}{\sqrt{n}}$$

9.45 $n = .25\left[\dfrac{1.96}{B}\right]^2 = .25\left[\dfrac{1.96}{.10}\right]^2 = 96.04$; take $n = 97$.

9.47 Change the z critical value from 1.96 to 2.58. Thus the formula would be

$$n = .25\left[\frac{2.58}{B}\right]^2$$

9.49 a) 2.12
     b) 1.80
     c) 2.81
     d) 1.71
     e) 1.78
     f) 2.26

9.51 The t-critical value for a 90% confidence interval when df = 10 - 1 = 9 is 1.83. From the given data, n = 10, Σx = 219, and Σx^2 = 4949.92. From the summary statistics,

$$\overline{x} = \frac{219}{10} = 21.9$$

$$s^2 = \frac{4949.92 - \frac{(219)^2}{10}}{9} = \frac{4949.92 - 4796.1}{9} = \frac{153.82}{9} = 17.09$$

$$s = \sqrt{17.09} = 4.134 .$$

The 90% confidence interval based on this sample data is

$$\overline{x} \pm (t \text{ critical}) \frac{s}{\sqrt{n}} \rightarrow 21.9 \pm (1.83) \frac{4.134}{\sqrt{10}}$$

$$\rightarrow 21.9 \pm 2.39 \rightarrow (19.51, 24.29) .$$

9.53 a) n = 16, $\overline{x}$ = 6.6, s = 3.2

$$\overline{x} \pm (t \text{ critical}) \frac{s}{\sqrt{n}} = 6.6 \pm (1.75) \frac{3.2}{\sqrt{16}}$$

$$= 6.6 \pm 1.4 = (5.2, 8.0)$$

b) n = 16, $\overline{x}$ = 40.3, s = 16

$$\overline{x} \pm (t \text{ critical}) \frac{s}{\sqrt{n}} = 40.3 \pm (2.13) \frac{16}{\sqrt{16}}$$

$$= 40.3 \pm 8.52 = (31.78, 48.82)$$

9.55 a) $$\overline{x} \pm (t \text{ critical}) \frac{s}{\sqrt{n}} = 428 \pm (2.13) \frac{35}{\sqrt{5}}$$

$$= 428 \pm 33.34 = (394.66, 461.34)$$

b) It would be narrower, since the standard deviation for the times from the food dehydrator is smaller than the standard deviation of times for the convection oven.

c) For convection oven:

$$2199 \pm (2.78) \frac{100}{\sqrt{5}} = 2199 \pm 124.33 = (2074.67, 2323.33)$$

For food dehydrator:

$$3920 \pm (2.78) \frac{170}{\sqrt{5}} = 3920 \pm 211.35 = (3708.65, 4131.35)$$

For microwave oven:

$$1431 \pm (2.78) \frac{30}{\sqrt{5}} = 1431 \pm 37.3 = (1393.7, 1468.3)$$

It appears that the microwave oven procedure is the most energy efficient, with convection oven being second best, while the food dehydrator method is the least energy efficient.

9.57 a) Summary statistics are:
$n = 9$, $\overline{x} = 263.7$, $s = 50.4$
The 90% confidence interval is:

$\overline{x} \pm$ (t critical) $\dfrac{s}{\sqrt{n}} = 263.7 \pm (1.86)\dfrac{50.4}{\sqrt{9}}$

$= 263.7 \pm 31.25 = (232.45, 294.95)$

b) Summary statistics are:
$n = 22$, $\overline{x} = 58.5$, $s = 12.1$
The 99% confidence interval is:

$\overline{x} \pm$ (t critical) $\dfrac{s}{\sqrt{n}} = 58.5 \pm (2.83)\dfrac{12.1}{\sqrt{22}}$

$= 58.5 \pm 7.3 = (51.2, 65.8)$

c) The distribution of the number of operations performed in one year must be (at least approximately) normal for each of the two populations.

9.59 Summary statistics for the sample are:
$n = 5$, $\overline{x} = 17$, $s = 9.03$
The 95% confidence interval is given by

$\overline{x} \pm$ (t critical) $\dfrac{s}{\sqrt{n}} = 17 \pm (2.78)\dfrac{9.03}{\sqrt{5}}$

$= 17 \pm 11.23 = (5.77, 28.23)$

9.61 The 99% confidence interval for the mean commuting distance based on this sample is

$$\bar{x} \pm (t \text{ critical}) \frac{s}{\sqrt{n}} = 10.9 \pm (2.58) \frac{6.2}{\sqrt{300}}$$

$$\rightarrow 10.9 \pm 0.924 \rightarrow (9.976, \ 11.824)$$

9.63 Let π denote the proportion of all VCR owners who use the fast-forward feature to avoid advertisements. The point estimate of π is p = 715/1100 = 0.65, and the 95% confidence interval is

$$.65 \pm 1.96\sqrt{\frac{.65(.35)}{1100}} = .65 \pm 1.96(.0144)$$

$$= .65 \pm .0282 = (.6218, \ .6782).$$

With 95% confidence, the true percentage of all VCR owners who use the fast-forward feature to avoid advertisements is estimated to be between 62% and 68%.

9.65 Let π denote the true proportion of all shoppers who purchase generic brands. The point estimate of π is p = 727/1442 = .5042 The 99% confidence interval for π is:

$$.5042 \pm (2.58)\sqrt{\frac{(.5042)(.4958)}{1442}} = .5042 \pm (2.58)(.0132)$$

$$= .5042 \pm .0340 = (.4702, \ .5382)$$

9.67 Let μ denote the mean concentration of tetracycline for all dogs, if they were to receive tetracycline (55 mg/kg body weight) daily. If the distribution of concentration of tetracycline is normal, then the 95% confidence interval for μ is:

$$\bar{x} \pm (t \text{ critical}) \frac{s}{\sqrt{n}} = 138 \pm (2.26) \frac{65}{\sqrt{10}}$$

$$= 138 \pm 46.45 = (91.55, \ 184.45)$$

9.69 Let π denote the true proportion of Americans that think autoworkers are overpaid. The point estimate of π is p = 606/1515 = 0.4. The 90% confidence interval for π is:

$$.4 \pm (1.645)\sqrt{\frac{.4(.6)}{1515}} = .4 \pm .02 = (.38, \ .42)$$

9.71 Let π denote the actual proportion of inmates who are Caucasian. The point estimate of π is p = 25/31 = .8065. The 90% confidence interval for π is:

$$.8065 \pm (1.645)\sqrt{\frac{(.8065)(.1935)}{31}} = .8065 \pm .1167$$

$$= (.6898, \ .9232)$$

With 90% confidence, the percentage of Caucasian inmates at this prison is estimated to be between 69% and 92.3%.

9.73 The width of the interval discussed in the text is:

$$(\bar{x} + (1.96)\frac{s}{\sqrt{n}}) - (\bar{x} - (1.96)\frac{s}{\sqrt{n}}) = 2(1.96)\frac{s}{\sqrt{n}} = 3.92\frac{s}{\sqrt{n}}$$

The width of the interval suggested in this problem is:

$$(\bar{x} + (1.75)\frac{s}{\sqrt{n}}) - (\bar{x} - (2.33)\frac{s}{\sqrt{n}}) = (1.75 + 2.33)\frac{s}{\sqrt{n}} = 4.08\frac{s}{\sqrt{n}}$$

Since this latter interval is wider (less precise) than the one discussed in the text, its use it not recommended.

9.75 From the data,

$$\bar{y} = \frac{300 + 720 + 526 + 200 + 127}{5} = \frac{1873}{5} = 374.6$$

$$\bar{x} = \frac{300 + 520 + 526 + 200 + 157}{5} = \frac{1703}{5} = 340.6$$

$$\bar{d} = \frac{0 + 200 + 0 + 0 + (-30)}{5} = \frac{170}{5} = 34.0$$

The corresponding point estimates are:
mean per unit statistic = 5000(340.6) = 1,703,000
difference statistic = 1,761,300 - 5000(34) = 1,591,300
ratio statistic = $1,761,300 \frac{340.6}{374.6} = 1,601,438.28$

10.1 $\overline{x}$ = 50 is not a legitimate hypothesis, because $\overline{x}$ is a
statistic, not a population characteristic. Hypotheses are
always expressed in terms of a population characteristic, not in
terms of a sample statistic.

10.3   If we use the hypothesis H_0: μ = 100 versus H_a: μ > 100, we are
taking the position that the welds do not meet specifications,
and hence, are not acceptable unless there is substantial
evidence to show that the welds are good (i.e. μ > 100). If we
use the hypothesis H_0: μ = 100 versus H_a: μ < 100, we initially
believe the welds to be acceptable, and hence, will be declared
unacceptable only if there is substantial evidence to show that
the welds are faulty. It seems clear that we would choose the
first set-up, which places the burden of proof on the welding
contractor to show that the welds meet specifications.

10.5   Since the administration has decided to make the change if it
can conclude that **more than** 60% of the faculty favor the change,
the appropriate hypotheses are

$$H_0: \pi = .6 \quad \text{versus} \quad H_a: \pi > .6$$

Then, if H_0 is rejected, it can be concluded that more than 60%
of the faculty favor a change.

10.7   If this well is the only source of water for me, then I would
like the water quality control board to test the hypotheses

$$H_0: \mu = 30 \quad \text{versus} \quad H_a: \mu > 30$$

Then, only if the null hypothesis is rejected (i.e. the water is
determined to be bad) would I have to search for an alternative
water source.

However, if there is a readily available water source, then I
would want the water quality control board to test the
hypotheses

$$H_0: \mu = 30 \quad \text{versus} \quad H_a: \mu < 30$$

Then the current water source would continue to be used only if
it is determined that it is safe.

10.9   Since the manufacturer is interested in detecting values of μ
which are less than 40, as well as values of μ which exceed 40,
the appropriate hypotheses are:

$$H_0: \mu = 40 \quad \text{versus} \quad H_a: \mu \neq 40$$

10.11   a)   They failed to reject the null hypothesis, because their conclusion "There is no evidence of increased risk of death due to cancer" for those living in areas with nuclear facilities is precisely what the null hypothesis states.

         b)   They would be making a type II error since a type I error is failing to reject the null hypothesis when the null is false.

         c)   Since the null hypothesis is the initially favored hypothesis and is presumed to be the case until it is determined to be false, if we fail to reject the null hypothesis, it is not proven to be true. There is just not sufficient evidence to refute its presumed truth.

10.13   A type I error involves concluding that the water being discharged from the power plant has a mean temperature in excess of 150° F when, in fact, the mean temperature is not greater than 150° F. A type II error is concluding that the mean temperature of water being discharged is 150° F or less when, in fact, the mean temperature is in excess of 150° F. I would consider a type II error to be the more serious. If we make a type II error, then damage to the river ecosystem will occur. Since it generally takes a long time to repair such damage, I would want to avoid this error. A type I error means that we require the power plant to take corrective action when it was not necessary to do so. However, since there are alternative power sources, the consequences of this error are basically financial in nature.

10.15   A type I error is deciding that more than 30% are willing to pay this price for this type of personal computer when in fact fewer than 30% are willing to do so. The consequence of making this type of error is to market the PC and then not be able to sell it at that price and hence make a reasonable profit.

        A type II error is deciding that 30% or fewer are willing to pay this price when in fact more than 30% are willing to do so. The consequence of making this type of error is that the company would either not market the PC or market it at a lower price. In either case, the manufacturer does not make the profit that could have been realized if they had marketed the PC at $3200.

10.17   a)   Since a high level of mercury content is of particular concern, and the purpose of the test is to determine if μ is in excess of 5, the appropriate set of hypotheses is

$$H_0: \mu = 5 \quad \text{versus} \quad H_a: \mu > 5.$$

         b)   A type I error is concluding that the mercury content is unacceptably high, when it is not. In the interest of public safety, choose the larger significance level of 0.1.

10.19 The test statistic is $z = \dfrac{\overline{x} - 400}{\dfrac{s}{\sqrt{50}}}$.

Since H_a has the form of H_a: μ > hypothesized value, the rejection region has the form of z > (z critical value). The value for z critical value may be found using Table II (Standard Normal Curve Areas).

a)   z > 2.33

b)   z > 1.645

c)   z > 1.28

d)   z > 1.13

10.21 Since we are performing a lower tail test, the rejection region is in the form of z < - (z critical value).

a)   z < -2.33

b)   z < -1.645

c)   z < -1.28

10.23 1.   Let μ denote the true average maximum penetration.

       2. H_0: μ = 50

       3. H_a: μ > 50

       4.   Test statistic: $z = \dfrac{\overline{x} - 50}{\dfrac{s}{\sqrt{n}}}$

       5.   Rejection region: z > 1.645

       6.   Computations: n = 35, $\overline{x}$ = 52.7, s = 4.8,
             $z = \dfrac{52.7 - 50}{\dfrac{4.8}{\sqrt{35}}} = 3.328$

       7.   Conclusion: Since 3.328 exceeds 1.645, the computed z falls in the rejection region. At level of significance .05, H_0 is rejected. The data provides support for the conclusion that the true average maximum penetration exceeds 50 mils.

10.25 1.   Population characteristic of interest: μ = average age of brides marrying for the first time in 1990.

       2. H_0: μ = 20.8

       3. H_a: μ > 20.8

4. Test statistic: $z = \dfrac{\overline{x} - 20.8}{\dfrac{s}{\sqrt{n}}}$

5. Rejection region: $z > 2.33$

6. $z = \dfrac{23.9 - 20.8}{\dfrac{6.4}{\sqrt{100}}} = \dfrac{3.10}{0.64} = 4.84$

7. Conclusion: Since 4.84 is greater than 2.33, the calculated z does fall in the rejection region and H_0 is rejected at the .01 level. The data supports the conclusion that the mean age of brides marrying for the first time in 1990 is larger than that in 1970.

10.27 1. Let μ denote the true average mileage of cars brought to the dealer for a 3000-mile checkup.

2. $H_0: \mu = 3000$

3. $H_a: \mu \neq 3000$

4. Test statistic: $z = \dfrac{\overline{x} - 3000}{\dfrac{s}{\sqrt{n}}}$

5. Rejection region: $z < -2.58$ or $z > 2.58$

6. $z = \dfrac{3208 - 3000}{\dfrac{273}{\sqrt{50}}} = \dfrac{208}{38.61} = 5.39$

7. Conclusion: Since 5.39 is greater than 2.58, the calculated z does fall in the rejection region and H_0 is rejected at the .01 level. The data supports the conclusion that the mean initial checkup mileage differs from the manufacturer's recommended value.

10.29 1. Let μ denote the mean litter weight of Giza white rabbits.

2. $H_0: \mu = 2000$

3. $H_a: \mu < 2000$

4. Test statistic: $z = \dfrac{\overline{x} - 2000}{\dfrac{s}{\sqrt{n}}}$

5. Rejection region: $z < -1.645$

6. Computations: $n = 59$, $\overline{x} = 1888$, $s = 99$,

$z = \dfrac{1888 - 2000}{\dfrac{99}{\sqrt{59}}} = -8.6898$

7.   Conclusion: Since -8.6898 is less than -1.645, the computed z falls in the rejection region. At level of significance .05, H_0 is rejected. The data provides sufficient evidence to conclude that the mean litter weight is less than 2000 grams for Giza white rabbits.

10.31   1.   Let μ denote the mean iron-deficiency score for soybean plants grown in this soil type.

2. H_0: $\mu = 4.0$

3. H_a: $\mu > 4.0$

4.   Test statistic: $z = \dfrac{\bar{x} - 4.0}{\dfrac{s}{\sqrt{n}}}$

5.   Rejection region: $z > 2.33$

6.   Computations: $n = 79$, $\bar{x} = 4.2$, $s = .6$,

$$z = \dfrac{4.2 - 4}{\dfrac{.6}{\sqrt{79}}} = 2.96$$

7.   Conclusion: Since 2.96 exceeds 2.33, the computed z falls in the rejection region. At level of significance .01, H_0 is rejected. The sample contains sufficient evidence to conclude that the mean iron-deficiency score exceeds 4.0 for soybean plants grown in this type of soil.

10.33   1.   Population characteristic of interest: μ = the average lateral recumbency time when Ketamine is administered under these conditions.

2. H_0: $\mu = 20$ minutes

3. H_a: $\mu < 20$ minutes

4.   Test statistic: $z = \dfrac{\bar{x} - 20}{\dfrac{s}{\sqrt{n}}}$

5.   Rejection region: $z < -1.28$

6.   Computations: $n = 73$, $\bar{x} = 18.86$, $s = 8.6$,

$$z = \dfrac{18.86 - 20}{\dfrac{8.6}{\sqrt{73}}} = -1.13$$

7.   Conclusion: Since -1.13 is not less than -1.28, the computed z does not fall into the rejection region. At level of significance .10, H_0 is not rejected. The data does not provide support for concluding that the average lateral recumbency time is less than 20 minutes when Ketamine is administered under these conditions.

10.35　1.　　Let μ denote the mean playing time of this manufacturer's audio tapes.

2.　　H_0: $\mu = 90$

3.　　H_a: $\mu < 90$

4.　　Test statistic: $z = \dfrac{\overline{x} - 90}{\dfrac{s}{\sqrt{n}}}$

5.　　For $\alpha = 0.05$, reject H_0 if $z < -1.645$.

6.　　For $n = 900$, $\overline{x} = 89.95$, $s = .3$,

$$z = \dfrac{89.95 - 90}{\dfrac{.3}{\sqrt{900}}} = -5$$

7.　　Since the computed z of -5 is less than the z critical of -1.645, the computed z falls in the rejection region. H_0 is rejected, and it is concluded that the mean playing time is less than 90 minutes. Since H_0 was rejected, the results are said to be statistically significant. However, 89.95 is only .05 minutes below the claimed value of 90. The fact that the mean playing time is less than the claimed value by so little leads one to conclude that the results may not be of much practical significance.

10.37 H_0 is rejected if P-value $\leq \alpha$. H_0 would be rejected for only the following pair:

d)   P-value = .084, α = 0.10

10.39   Since this is a two-tailed test, the P-value is equal to twice the area captured in the tail in which z falls. Using Appendix Table II, the P-values are:

a)   2(.0179) = .0358

b)   2(.0446) = .0892

c)   2(.3085) = .6170

d)   2(.0808) = .1616

e)   2(0) = 0

10.41   a) $z = \dfrac{12.1 - 12}{\frac{.2}{\sqrt{36}}} = 3.0$

b)   Since this is an upper-tail test, the P-value equals the area under the z curve to the right of 3.0, which is .0013.

c) H_0 is rejected if the P-value is less than α. Since the P-value of .0013 is less than α, which is .05, H_0 would be rejected in favor of the conclusion that the machine is overfilling.

10.43   Population characteristic of interest: μ = average drying time when additive is used in the paint

H_0: μ = 75 H_a: μ < 75

Test statistic: $z = \dfrac{\overline{x} - 75}{\frac{s}{\sqrt{n}}}$

Computations: n = 100, $\overline{x}$ = 68.5, s = 9.4,

$z = \dfrac{68.5 - 75}{\frac{9.4}{\sqrt{100}}} = -6.91$

P-value = P(z < -6.91) < .0002

Conclusion: Since the P-value of .0002 is less than the chosen α of .01, H_0 is rejected and the experimental evidence indicates strongly that the additive shortens the drying time.

10.45   a)   Population characteristic of interest: μ = mean score on this test when students are given an "organizer".

H_0: $\mu = 32$ H_a: $\mu > 32$

Test statistic: $z = \dfrac{\overline{x} - 32}{\dfrac{s}{\sqrt{n}}}$

For $\alpha = .05$, reject H_0 if $z > 1.645$

Computations: $n = 100$, $\overline{x} = 32.96$, $s = 9.24$,

$z = \dfrac{32.96 - 32}{\dfrac{9.24}{\sqrt{100}}} = 1.04$

Conclusion: Since the calculated z of 1.04 does not exceed the z critical value of 1.645, H_0 is not rejected using a level of significance of .05. The data does not support the conclusion that preliminary exposure to an organizer increases the true average score on this test.

b)   P-value = $P(1.04 < z) = 1 - P(z \leq 1.04)$
          $= 1 - .8505 = .1492$
The null hypothesis would be rejected for significance levels that are .1492 or larger.

10.47

	Test Statistic	d.f.	Rejection Region
a)	$t = \dfrac{\bar{x} - 30}{\frac{s}{\sqrt{10}}}$	9	Reject H_0 if $t > 2.26$ or $t < -2.26$
b)	$t = \dfrac{\bar{x} - 30}{\frac{s}{\sqrt{18}}}$	17	Reject H_0 if $t > 2.90$ or $t < -2.90$
c)	$t = \dfrac{\bar{x} - 30}{\frac{s}{\sqrt{25}}}$	24	Reject H_0 if $t > 3.75$ or $t < -3.75$
d)	$t = \dfrac{\bar{x} - 30}{\frac{s}{\sqrt{50}}}$	49	Reject H_0 if $t > 1.67$ or $t < -1.67$

10.49 Let μ denote the true average activation time to first sprinkler activation using aqueous film-forming foam.

H_0: $\mu \leq 25$ $\qquad$ H_a: $\mu > 25$

$$t = \dfrac{\bar{x} - 25}{\frac{s}{\sqrt{n}}} \qquad \text{with d.f.} = 12$$

For $\alpha = 0.05$, reject H_0 if $t > 1.78$.

From the data: $n = 13$, $\bar{x} = 27.92$, $s = 5.62$,

$$t = \dfrac{27.92 - 25}{\frac{5.62}{\sqrt{13}}} = 1.875$$

Since the calculated t of 1.875 exceeds the t critical of 1.78, H_0 is rejected. Thus, at .05 level of significance, it is concluded that the average time to first sprinkler activation exceeds 25 seconds, and the design specifications have not been met.

The assumption being made is that activation time has a normal distribution.

10.51 Let μ = mean heat flux level when coal dust is used.

H_0: $\mu = 29.0$ $\qquad$ H_a: $\mu > 29.0$

Test statistic: $t = \dfrac{\bar{x} - 29.0}{\frac{s}{\sqrt{n}}} \qquad \text{with d.f.} = 8 - 1 = 7$

Rejection region: Reject H_0 if either $t > 1.90$.

From the data: n = 8, $\overline{x}$ = 30.7875, s = 6.53,

$$t = \frac{30.7875 - 29}{\frac{6.53}{\sqrt{8}}} = .77$$

t = .77 does not fall into the rejection region, so at level of significance .05, H_0 is not rejected. The sample data suggests that the mean heat flux level when coal dust is used is not greater than when grass is used.

10.53   Let μ denote the average heart rate for pigs under this anesthetic.

H_0: μ = 114 H_a: $\mu \neq$ 114

$$t = \frac{\overline{x} - 114}{\frac{s}{\sqrt{4}}} \qquad \text{with d.f. = 3}$$

For α = 0.10, reject H_0 if t > 2.35 or if t < -2.35.

From the data: n = 4, $\overline{x}$ = 109.25, s = 16.19,

$$t = \frac{109.25 - 114}{\frac{16.19}{\sqrt{4}}} = -0.59$$

Since the calculated t of -0.59 is between the t critical values of -2.35 and 2.35, H_0 is not rejected. There is not sufficient evidence to conclude that the mean heart rate while under the anesthetic differs from the normal mean heart rate.

10.55   Let μ denote the true average rate of uptake of radio-labeled amino acid when grown in a medium containing nitrates.

H_0: μ = 8000 H_a: μ < 8000

Assuming that rate of uptake has a normal distribution, the test statistic is

$$t = \frac{\overline{x} - 8000}{\frac{s}{\sqrt{n}}} \qquad \text{with d.f. = 14}$$

For α = 0.10, reject H_0 if t < -1.35.

From the data given, n = 15, $\overline{x}$ = 7788.8, s = 1002.4308,

$$t = \frac{7788.8 - 8000}{\frac{1002.4308}{\sqrt{15}}} = -0.816$$

Since the calculated t value of -0.816 is not less than -1.35, the null hypothesis is not rejected. It cannot be concluded that the addition of nitrates results in a decrease in true average uptake.

10.57   a)   .10 > P-value > .05

        b)   .01 > P-value > .002

        c)   .001 > P-value

        d)   P-value > .20

10.59   From the solution to 10.49, the computed t = 1.875 with d.f. = 12. From the t-table, it is found that .05 > P-value > .025. Thus at the .01 level of significance, H_0 would not be rejected. The conclusion would be that the design specifications have been met.

10.61   1.   Let π denote the proportion of all pizzas ordered that are large.

2. H_0: π = .75

3. H_a: π > .75

4.   Test statistic: $z = \dfrac{p - .75}{\sqrt{\dfrac{.75(.25)}{n}}}$

5.   Rejection region: z > 1.28

6.   Computations: n = 150, p = $\dfrac{120}{150}$ = .80,

$z = \dfrac{.80 - .75}{\sqrt{\dfrac{.75(.25)}{150}}}$ = 1.41

7.   Since 1.41 exceeds 1.28, the computed z falls in the rejection region. At level of significance .10, H_0 is rejected. The sample provides sufficient evidence to conclude that the true proportion of pizzas ordered that are of the large size exceeds .75.

10.63   1.   Let π represent the proportion of customers who favor underground installation.

2. H_0: π = .60

3. H_a: π > .60

4.   Since $n\pi$ = 160(.60) = 96 $\geq$ 5, and n(1-π) = 160(.4) = 64 $\geq$ 5, the large sample z test may be used.

$z = \dfrac{p - .6}{\sqrt{\dfrac{.6(.4)}{160}}} = \dfrac{p - .6}{.0387}$

5.   For α = .05, reject H_0 if z > 1.645.

6.   p = $\dfrac{118}{160}$ = .7375

$z = \dfrac{.7375 - .6}{.0387}$ = 3.55

7.   Since the calculated z of 3.55 exceeds the z critical value of 1.645, H_0 is rejected. It is concluded that the proportion of all customers who favor underground installation, in spite of the surcharge, exceeds 60%, and hence, the company should do the installation underground.

10.65   1.   Let π represent the proportion of job applicants in California that test positive for drug use.

2. H_0: π = .10

3. H_a: $\pi > .10$

4.   Since $n\pi = 600(.10) = 60 \geq 5$, and $n(1-\pi) = 600(.90) = 540 \geq 5$, the large sample z test may be used.

$$z = \frac{p - .1}{\sqrt{\frac{.1(.9)}{n}}}$$

5.   For $\alpha = .01$, reject H_0 if $z > 2.33$

6. $p = \frac{73}{600} = .1217$

$$z = \frac{.1217 - .1}{\sqrt{\frac{.1(.9)}{600}}} = \frac{.0217}{.0122} = 1.78$$

7.   Since the calculated z of 1.78 does not exceed the z critical value of 2.33, H_0 is not rejected. At level of significance .01, the data does not support the conclusion that the proportion of job applicants in California that test positive for drug use exceeds .10.

10.67   1.   Let π denote the proportion of weapons that would not be detected.

2. H_0: $\pi = .15$

3. H_a: $\pi > .15$

4.   Test statistic: $z = \frac{p - .15}{\sqrt{\frac{.15(.85)}{n}}}$

5.   Rejection region: $z > 2.33$

6.   Computations: $n = 2419$, $p = \frac{496}{2419} = .205$,

$$z = \frac{.205 - .15}{\sqrt{\frac{.15(.85)}{2419}}} = 7.58$$

7.   Since 7.58 exceeds 2.33, the computed z falls into the rejection region. At level of significance .01, H_0 is rejected. The sample provides sufficient evidence to conclude that the proportion of weapons that would not be detected exceeds .15.

10.69   1.   Let π denote the proportion of westerners who have attended at least one year of college.

2. H_0: $\pi = .32$

3. H_a: $\pi > .32$

4.   Test statistic: $z = \dfrac{p - .32}{\sqrt{\dfrac{.32(.68)}{n}}}$

5.   Rejection region: $z > 2.33$

6.   Computations: $n = 200$, $p = \dfrac{82}{200} = .41$,

    $z = \dfrac{.41 - .32}{\sqrt{\dfrac{.32(.68)}{200}}} = 2.73$

7.   Since 2.73 exceeds 2.33, the computed z falls in the rejection region. At level of significance .01, H_0 is rejected. The sample provides sufficient evidence to conclude that the proportion of westerners who have attended at least one year of college is greater than that for the U.S. as a whole.

10.71   1.   Let π denote the true proportion of Californians who feel that their life is stressful.

      2. H_0: $\pi = .5$

      3. H_a: $\pi < .5$

      4.   Test statistic: $z = \dfrac{p - .5}{\sqrt{\dfrac{.5(.5)}{n}}}$

      5.   Rejection region: $z < -1.645$

      6.   Computations: $n = 1008$, $p = \dfrac{484}{1008} = .4802$

      $z = \dfrac{.4802 - .5}{\sqrt{\dfrac{.5(.5)}{1008}}} = -1.26$

      7.   Since -1.26 is not less than -1.645, the computed z does not fall into the rejection region. At level of significance .05, H_0 is not rejected. The sample does not contain sufficient evidence to conclude that fewer than half of Californians feel that their life is stressful.

10.73   a)   1.   Let π represent the miscarriage rate of women who work full time on VDT's.

          2. H_0: $\pi = .10$

          3. H_a: $\pi > .10$

          4.   Since $n\pi = 48(.1) = 4.8 \approx 5 \geq 5$, and $n(1-\pi) = 48(.9) = 43.2 \geq 5$, the large sample z test for π may be used.

$$z = \frac{p - .1}{\sqrt{\frac{.1(.9)}{48}}} = \frac{p - .1}{.0433}$$

5.   For $\alpha = 0.01$, reject H_0 if $z > 2.33$.

6.   From the data, $p = \frac{15}{48} = .3125$

$$z = \frac{3125 - .1}{.0433} = \frac{.2125}{.0433} = 4.91$$

7.   Since the z calculated value 4.91 exceeds the z
        critical value of 2.33, H_0 is rejected. It can be
        concluded that the miscarriage rate of women who
        work full time on VDT's is higher than that of the
        general population.

b)   No. it cannot be concluded that the full time work on VDT
        tends to **cause** miscarriages. The above hypothesis test
        indicates only that the rate of miscarriages for women who
        work full time on VDT's is higher than for those who do
        not work full time on VDT's. It does not imply a causal
        effect between full time work on VDT's and miscarriage
        rate. For example, it may be that prolonged sitting is the
        cause of the increased incidence rate, rather than the
        VDT's. It may be that any occupation with prolonged
        sitting would have a higher than usual incidence rate for
        miscarriage.

10.75   a)   The z statistic given in this problem is an appropriate test statistic in this setting because:

     i)   The parameter being tested is a population mean.
    ii)   The variance of the population being sampled is assumed known.
   iii)   The sample size is sufficiently large, and hence by the central limit theorem, the distribution of the random variable $\overline{x}$ should be approximately normal.

     b)   A type I error involves concluding that the water being discharged from the power plant has a mean temperature in excess of 150° F when, in fact, the mean temperature is not greater than 150° F. A type II error is concluding that the mean temperature of water being discharged is 150° F or less when, in fact, the mean temperature is in excess of 150° F.

     c)   From Appendix Table II, the area to the right of 1.8 is .0359. Hence, rejecting H_0 when z > 1.8 corresponds to an α value of 0.0359.

     d)   If z > 1.8, then $\dfrac{\overline{x} - 150}{\frac{10}{\sqrt{50}}} > 1.8,,$

and it follows that $\overline{x} > 150 + 1.8\left(\dfrac{10}{\sqrt{50}}\right) = 152.55.$

In the graph below,
the shaded area = P(type II error when μ = 153)

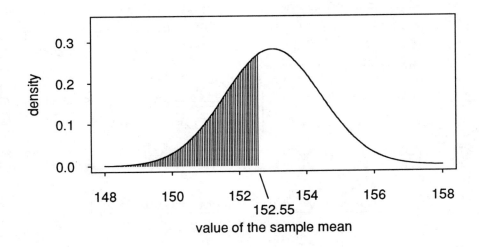

e) β (when $\mu = 153$) $= P(\overline{x} < 152.55) = P\left[z < \dfrac{152.55 - 153}{\dfrac{10}{\sqrt{50}}}\right]$

$= P(z < -.32) = .3745$

f) β (when $\mu = 160$) $= P(\overline{x} < 152.55) = P\left[z < \dfrac{152.55 - 160}{\dfrac{10}{\sqrt{50}}}\right]$

$= P(z < -5.27) \approx 0$

g) From part (d), H_0 is rejected if $\overline{x} > 152.55$. Since $\overline{x} = 152.4$, H_0 is not rejected. Because H_0 is not rejected, a type II error might have been made.

10.77 a) Let π be the true proportion of apartments which prohibit children.

H_0: $\pi = .75$ $\qquad$ H_a: $\pi > .75$

Since $n\pi = 125(.75) = 93.75 \geq 5$, and $n(1-\pi) = 125(.25) = 31.25 \geq 5$, the large sample z test for π may be used.

For $\alpha = .05$, reject H_0 if $z > 1.645$.

$z = \dfrac{p - .75}{\sqrt{\dfrac{.75(.25)}{125}}} = \dfrac{.816 - .75}{.0387} = 1.71$

Since the calculated z of 1.71 exceeds the z critical value of 1.645, H_0 is rejected. This .05 level test does lead to the conclusion that more than 75% of the apartments exclude children.

b) The test with $\alpha = .05$ reject H_0 if $\dfrac{p - .75}{.0387} > 1.645$, which is equivalent to $p > .75 + .0387(1.645) = .8137$. H_0 will then not be rejected if $p \leq .8137$.

When $\pi = .80$ and $n = 125$,
$\beta = P(\text{not rejecting } H_0 \text{ when } \pi = .8)$
$= P(p \leq .8137) = P\left[z \leq \dfrac{.8137 - .8}{\sqrt{\dfrac{.8(.2)}{125}}}\right] = P(z \leq .38) = .6480.$

10.79 a) Since this is an upper-tail test with $\alpha = 0.05$ and d.f. $= 9$, the rejection region would be t values which exceed 1.83.

b) $d = \dfrac{|\text{alternative value} - \text{hypothesized value}|}{\sigma}$

i) $d = \dfrac{|52 - 50|}{10} = .2$
From Appendix Table V, $\beta = .85$.

ii) $d = \dfrac{|55 - 50|}{10} = .5$

From Appendix Table V, $\beta = .55$.

iii) $d = \dfrac{|60 - 50|}{10} = 1$

From Appendix Table V, $\beta = .10$.

iv) $d = \dfrac{|70 - 50|}{10} = .5$

From Appendix Table V, $\beta \approx 0$.

10.81   Let π denote the proportion of all males with birth weight of
2500 grams or less who are unfit for military service.

$H_0: \pi = .062$ $H_a: \pi > .062$

The test statistic is: $z = \dfrac{p - .062}{\sqrt{\dfrac{(.062)(.938)}{105}}}$.

For $\alpha = 0.01$, reject H_0 if $z > 2.33$.

From the sample: $p = \dfrac{7}{105} = .0667$

$z = \dfrac{.0667 - .062}{\sqrt{\dfrac{(.062)(.938)}{105}}} = \dfrac{0.0047}{0.0235} = 0.2$

Since the calculated z of 0.2 does not exceed the z critical
value of 2.33, the null hypothesis is not rejected. The data
does not provide sufficient evidence to indicate that the true
proportion of males with birth weight of 2500 g. or less, who
are unfit for military service, is higher than that of the
general population.

10.83   a)   Let π represent the response rate when the distributor is
stigmatized by an eye patch.

$H_0: \pi = .40$ $H_a: \pi > .40$

The test statistic is: $z = \dfrac{p - .40}{\sqrt{\dfrac{(.4)(.60)}{n}}}$.

For $\alpha = 0.05$, reject H_0 if $z > 1.645$.

From the data: $n = 200$, $p = \dfrac{109}{200} = .545$,

$z = \dfrac{.545 - .40}{\sqrt{\dfrac{(.40)(.60)}{200}}} = 4.19$

Since the calculated z of 4.19 exceeds the z critical
value of 1.645, the null hypothesis is rejected. The data
strongly suggests that the response rate does exceed the
rate in the past.

b)   P-value = $P(z > 4.19) < .0002$
Since the P-value of .0002 is less than the α value of
.05, H_0 is rejected.

10.85   Let π denote the proportion of senior citizens who are satisfied
with their grocery purchases.

$H_0: \pi = .8$ $H_a: \pi < .8$

The test statistic is: $z = \dfrac{p - .8}{\sqrt{\dfrac{(.8)(.2)}{n}}}$.

For $\alpha = .01$, reject H_0 if $z < -2.33$.

From the sample: $p = \dfrac{270}{404} = .6683$,

$z = \dfrac{.6683 - .8}{\sqrt{\dfrac{(.8)(.2)}{404}}} = -6.62$

Since the calculated z of -6.62 is less than the z critical of -2.33, the null hypothesis is rejected. The data strongly suggests that the proportion of senior citizens who are satisfied with their grocery purchases is smaller than that for the general population.

10.87   Let μ denote the mean met-enkephalin level of SIDS victims.

H_0: $\mu = 7.48$ H_a: $\mu > 7.48$

The test statistic is: $t = \dfrac{\overline{x} - 7.48}{\dfrac{s}{\sqrt{n}}}$ with d.f. = 11.

For $\alpha = .05$, reject H_0 if $t > 1.80$.

Using the information from the sample,

$t = \dfrac{7.66 - 7.48}{\dfrac{3.78}{\sqrt{12}}} = 0.16$.

Since the t calculated value of 0.16 does not fall in the rejection region, the null hypothesis is not rejected. The data does not allow one to conclude that mean met-enkephalin level of SIDS victims is higher than that of children not suffering from SIDS.

10.89   Let μ denote the mean root length of pearl millet when irrigated with the 50% wastewater solution.

H_0: $\mu = 6.40$ H_a: $\mu \neq 6.40$

The test statistic is: $z = \dfrac{\overline{x} - 6.40}{\dfrac{s}{\sqrt{n}}}$.

For $\alpha = .05$, reject H_0 if $z < -1.96$ or $z > 1.96$.

Using the sample data: $z = \dfrac{4.76 - 6.4}{\dfrac{.48}{\sqrt{40}}} = -21.6$.

The calculated z value of -21.6 falls in the rejection region, hence, the null hypothesis is rejected. The data strongly suggests that irrigation with the wastewater solution results in a mean root length that differs from 6.40.

10.91 a) Let μ denote the true mean time required to achieve 100° F with the heating equipment of this manufacturer.

$H_0: \mu = 15$ $H_a: \mu > 15$

The test statistic is: $z = \dfrac{\overline{x} - 15}{\dfrac{s}{\sqrt{n}}}$.

For $\alpha = 0.05$, reject H_0 if $z > 1.645$.

From the sample: $n = 32$, $\overline{x} = 17.5$, $s = 2.2$,

$z = \dfrac{17.5 - 15}{\dfrac{2.2}{\sqrt{32}}} = 6.43$.

The calculated z value of 6.43 falls in the rejection region, the null hypothesis is rejected. The data does cast doubt on the company's claim that it requires at most 15 minutes to achieve 100° F.

b) P-value = $P(z > 6.43) < .0002$
Because the P-value is smaller than α, H_0 is rejected.

CHAPTER 11
INFERENCES USING TWO INDEPENDENT SAMPLES

Section 1

11.1　$\mu_{\bar{x}_1 - \bar{x}_2} = \mu_1 - \mu_2 = 30 - 25 = 5$

$$\sigma_{\bar{x}_1 - \bar{x}_2} = \sqrt{\frac{\sigma_1^2}{n_1} + \frac{\sigma_2^2}{n_2}} = \sqrt{\frac{(2)^2}{40} + \frac{(3)^2}{50}} = \sqrt{\frac{4}{40} + \frac{9}{50}}$$
$$= \sqrt{.28} = .529$$

Since both n_1 and n_2 are large, the sampling distribution of $\bar{x}_1 - \bar{x}_2$ is approximately normal. It is centered at 5 and the standard deviation is .529.

11.3　Let μ_1 and μ_2 denote the true average counts for two-year and four-year old horses, respectively.

$H_0: \mu_1 - \mu_2 = 0$　　　$H_a: \mu_1 - \mu_2 < 0$

$$z = \frac{(\bar{x}_1 - \bar{x}_2) - 0}{\sqrt{\frac{s_1^2}{n_1} + \frac{s_2^2}{n_2}}}$$

For $\alpha = 0.001$, reject H_0 if $z < -3.09$.

$$z = \frac{(51 - 56) - 0}{\sqrt{\frac{(5.6)^2}{197} + \frac{(4.3)^2}{77}}} = \frac{-5}{\sqrt{.4}} = \frac{-5}{.63} = -7.91$$

Since the calculated z of -7.91 is less than the z critical value -3.09, H_0 is rejected. The data strongly suggests that the average neutrophil count for four-year olds exceeds that for two-year olds.

11.5　Let μ_1 denote the true mean approval rating for male players and μ_2 the true mean approval rating for female players.

$H_0: \mu_1 - \mu_2 = 0$　　　$H_a: \mu_1 - \mu_2 > 0$

$$z = \frac{(\bar{x}_1 - \bar{x}_2) - 0}{\sqrt{\frac{s_1^2}{n_1} + \frac{s_2^2}{n_2}}}$$

For $\alpha = 0.05$, reject H_0 if $z > 1.645$.

$$z = \frac{(2.76 - 2.02) - 0}{\sqrt{\frac{(.44)^2}{56} + \frac{(.41)^2}{67}}} = \frac{0.74}{0.0772} = 9.58$$

Since the computed z of 9.58 exceeds the critical z of 1.645, the null hypothesis is rejected. At level of significance 0.05, the data supports the conclusion that the mean approval rating is higher for males than for females.

11.7   Let μ_1 and μ_2 denote the true mean number of friends consulted prior to purchase of a VCR and a large TV, respectively.

$H_0: \mu_1 - \mu_2 = 0$ $\qquad$ $H_a: \mu_1 - \mu_2 > 0$

$$z = \frac{(\overline{x}_1 - \overline{x}_2) - 0}{\sqrt{\dfrac{s_1^2}{n_1} + \dfrac{s_2^2}{n_2}}}$$

For $\alpha = 0.01$, reject H_0 if $z > 2.33$.

$$z = \frac{(3.26 - 1.65) - 0}{\sqrt{\dfrac{(.6)^2}{50} + \dfrac{(.4)^2}{50}}} = \frac{1.61}{.1020} = 15.78$$

Since the computed z of 15.78 is greater than the critical z of 2.33, H_0 is rejected. The data suggests very strongly that the average number of friends consulted prior to purchase is higher for those purchasing a VCR than for those purchasing a large TV.

11.9   a) $$(\overline{x}_1 - \overline{x}_2) \pm 1.645 \sqrt{\frac{s_1^2}{n_1} + \frac{s_2^2}{n_2}}$$

$$(93 - 92) \pm 1.645 \sqrt{\frac{(9)^2}{74} + \frac{(8)^2}{72}}$$

$= 1 \pm 1.645 \sqrt{1.983} = 1 \pm 1.645 \, (1.408) = 1 \pm 2.32$
$= (-1.32, \ 3.32)$

b)   Let μ_1 and μ_2 denote the true mean systolic blood pressure of 1 year-old males and 3 year-old males, respectively.

$H_0: \mu_1 - \mu_2 = 0$ $\qquad$ $H_a: \mu_1 - \mu_2 \neq 0$

$$z = \frac{(\overline{x}_1 - \overline{x}_2) - 0}{\sqrt{\dfrac{s_1^2}{n_1} + \dfrac{s_2^2}{n_2}}}$$

For $\alpha = 0.01$, reject H_0 if $z > 1.96$ or $z < -1.96$.

$$z = \frac{(93 - 96) - 0}{\sqrt{\dfrac{(9)^2}{74} + \dfrac{(10)^2}{86}}} = \frac{-3}{1.5} = -2.00$$

Since the computed z of -2.00 is less than the critical z of -1.96, H_0 is rejected. The data contains sufficient evidence to conclude that there is a difference in the mean systolic blood pressures of 1 year-old and 3 year-old males.

c) Let μ_1 and μ_2 denote the true mean systolic blood pressure of 1 year-old females and 3 year-old females, respectively.

$$H_0: \mu_1 - \mu_2 = 0 \qquad H_a: \mu_1 - \mu_2 \neq 0$$

$$z = \frac{(\overline{x}_1 - \overline{x}_2) - 0}{\sqrt{\dfrac{s_1^2}{n_1} + \dfrac{s_2^2}{n_2}}}$$

For $\alpha = 0.01$, reject H_0 if $z > 1.96$ or $z < -1.96$.

$$z = \frac{(92 - 96) - 0}{\sqrt{\dfrac{(8)^2}{72} + \dfrac{(10)^2}{66}}} = \frac{-4}{1.55} = -2.58$$

Since the computed z of -2.58 is less than the critical z of -1.96, H_0 is rejected. The data contains sufficient evidence to conclude that there is a difference in the mean systolic blood pressures of 1 year-old and 3 year-old females.

11.11 The 99% confidence interval for $\mu_1 - \mu_2$ is:

$$(\overline{x}_1 - \overline{x}_2) \pm 2.58 \sqrt{\frac{s_1^2}{n_1} + \frac{s_2^2}{n_2}}$$

which becomes:

$$(33.40 - 35.39) \pm 2.58 \sqrt{\frac{(.428)^2}{51} + \frac{(.294)^2}{54}}$$

$$-1.99 \pm (2.58) \sqrt{.0052} = -1.99 \pm (2.58)(.0721) = -1.99 \pm .186$$

Therefore, with 99% confidence $\mu_1 - \mu_2$ is estimated to be between -2.176 and -1.804.

It is not necessary to make any assumptions about the two salinity distributions. Even if they are not normal, the distribution of $\overline{X}_1 - \overline{X}_2$ will be approximately normal because of the large sample sizes.

11.13 a) Let μ_1 denote the true mean comprehension score for students hearing a time-compressed lecture and μ_2 the true mean comprehension score for students who hear a lecture at normal speed.

$$H_0: \mu_1 - \mu_2 = 0 \qquad H_a: \mu_1 - \mu_2 < 0$$

$$z = \frac{(\overline{x}_1 - \overline{x}_2) - 0}{\sqrt{\dfrac{s_1^2}{n_1} + \dfrac{s_2^2}{n_2}}}$$

For $\alpha = 0.01$, reject H_0 if $z < -2.33$.

$$z = \frac{(6.34 - 9.18) - 0}{\sqrt{\dfrac{(4.93)^2}{50} + \dfrac{(4.59)^2}{50}}} = \frac{-2.84}{\sqrt{.4861 + .4214}} = \frac{-2.84}{.9526} = -2.98$$

Since the calculated z value of -2.98 is less than the z critical value of -2.33, H_0 is rejected. The data strongly suggests that the mean comprehension score for students hearing a time-compressed lecture is lower than the mean score for students hearing a lecture at the normal speed.

b)   A 95% confidence level for $\mu_1 - \mu_2$ is:

$$(\overline{x}_1 - \overline{x}_2) \pm 1.96 \sqrt{\frac{s_1^2}{n_1} + \frac{s_2^2}{n_2}}$$

$$(6.34 - 9.18) \pm 1.96 \sqrt{\frac{(4.93)^2}{50} + \frac{(4.59)^2}{50}}$$

$$-2.84 \pm (1.96)(.9526) = -2.84 \pm 1.87 = (-4.71, -0.97)$$

The 95% confidence interval for $\mu_1 - \mu_2$ contains the values from -4.71 to -0.97. Since both values are negative, this indicates that the mean score for students hearing a time-compressed lecture is lower than the mean score for students hearing a lecture at normal speed. The difference may be as much as 4.71 points lower to as little as 0.97 points lower.

11.15 a)   Let μ_1 denote the mean salary for males and μ_2 the mean salary for females.

H_0: $\mu_1 - \mu_2 = 0$ H_a: $\mu_1 - \mu_2 > 0$

$$z = \frac{(\overline{x}_1 - \overline{x}_2) - 0}{\sqrt{\dfrac{s_1^2}{n_1} + \dfrac{s_2^2}{n_2}}}$$

For $\alpha = 0.05$, reject H_0 if $z > 1.645$.

$$z = \frac{(1634.10 - 1091.80) - 0}{\sqrt{\dfrac{(715)^2}{50} + \dfrac{(418.8)^2}{50}}} = \frac{542.3}{\sqrt{10224.5 + 3507.8}} = \frac{542.3}{117.19}$$

$z = 4.63$

Since the calculated z value exceeds the z critical value 1.645, H_0 is rejected. The data very strongly suggests that the mean salary for male college professors is greater than the mean salary for female college professors. This conclusion does not by itself point to discrimination against female professors. Some possible non-discriminating explanations for the observed differences are:

1. The amount of experience of a professor.

2. The rank of the professor.

3. Field of specialization (relatively few females are in engineering or business which are high paying disciplines; many females are in the humanities and social sciences, where pay is not as high.)

b) P-value = P(z > 4.63) < .0002

c) Let μ_1 denote the true mean number of years at the university for males and μ_2 the true mean number of years at the university for females. A 90% confidence interval for $\mu_1 - \mu_2$ is:

$$(\overline{x}_1 - \overline{x}_2) \pm 1.645 \sqrt{\frac{s_1^2}{n_1} + \frac{s_2^2}{n_2}}$$

$$(7.93 - 6.25) \pm 1.645 \sqrt{\frac{(8.04)^2}{50} + \frac{(7.65)^2}{50}}$$

$$1.68 \pm (1.645)(1.569) = 1.68 \pm 2.58$$

The 90% confidence interval for the difference in mean years at the university for male and female college professors is (-0.9, 4.26). Since this interval contains the value zero, the null hypothesis of equal mean time at the university for male and female professors would not be rejected at level of significance .10. There does not appear to be sufficient evidence to conclude that the mean time differs for the two sexes.

11.17 Let μ_1 denote the true mean believability rating when a celebrity is used. Let μ_2 denote the true mean believability rating when an unknown actor is used.

$H_0: \mu_1 - \mu_2 = 0$ $\qquad$ $H_a: \mu_1 - \mu_2 \neq 0$

$$z = \frac{(\overline{x}_1 - \overline{x}_2) - 0}{\sqrt{\frac{s_1^2}{n_1} + \frac{s_2^2}{n_2}}}$$

For $\alpha = 0.05$, reject H_0 if $z > 1.96$ or $z < -1.96$.

$$z = \frac{(3.82 - 3.97) - 0}{\sqrt{\dfrac{(2.63)^2}{98} + \dfrac{(2.51)^2}{98}}} = \frac{-.15}{\sqrt{.0706 + .0643}} = \frac{-.15}{.3672} = -.41$$

Since the calculated z value of -.41 falls between the two z critical values -1.96 and 1.96, H_0 is not rejected. There is not sufficient evidence to conclude that the use of a celebrity results in a different mean believability rating than when an unknown actor is used.

11.19   Let $\mu_1 - \mu_2$ denote the true difference in mean stopping distances using disk brakes and using pneumatic brakes for cars of this type.

H_0: $\mu_1 - \mu_2 = -10$ H_a: $\mu_1 - \mu_2 < -10$

$$t = \frac{(\overline{x}_1 - \overline{x}_2) - (-10)}{\sqrt{s_p^2\left(\frac{1}{n_1} + \frac{1}{n_2}\right)}} \qquad \text{with d.f.} = 6 + 6 - 2 = 10$$

For $\alpha = 0.01$, reject H_0 if $t < -2.76$.

$$s_p^2 = \left(\frac{5}{10}\right)(5.03)^2 + \left(\frac{5}{10}\right)(5.38)^2$$

$$= \left(\frac{1}{2}\right)(25.30) + \left(\frac{1}{2}\right)(28.94) = 27.12$$

$$t = \frac{(115.7 - 129.3) + 10}{\sqrt{27.12\left(\frac{1}{6} + \frac{1}{6}\right)}} = \frac{-3.6}{\sqrt{9.04}} = \frac{-3.6}{3.01} = -1.20$$

Since the t calculated of -1.20 is greater than -2.76, H_0 is not rejected. There is insufficient evidence to conclude that the difference in mean stopping distances using disk brakes and pneumatic brakes is less than -10.

11.21   Let μ_1 denote the mean tree density for fertilized plots. Let μ_2 denote the mean tree density for non-fertilized plots.

H_0: $\mu_1 - \mu_2 = 0$ H_a: $\mu_1 - \mu_2 \neq 0$

$$t = \frac{(\overline{x}_1 - \overline{x}_2) - 0}{\sqrt{s_p^2\left(\frac{1}{n_1} + \frac{1}{n_2}\right)}} \qquad \text{with d.f.} = 8 + 8 - 2 = 14$$

For $\alpha = 0.05$, reject H_0 if $t < -2.15$ or if $t > 2.15$.

From the sample data for fertilized plots:
n = 8, $\Sigma X = 9472$, $\Sigma X^2 = 11,325,440$. From these,
$\overline{x} = 1184$ and $s^2 = 15,798.8571$

From the sample data for non-fertilized (control) plots:
n = 8, $\Sigma X = 9568$, $\Sigma X^2 = 11,541,504$. From these,
$\overline{x} = 1196$ and $s^2 = 14,025.1429$

$$s_p^2 = \left(\frac{8}{16}\right)(15798.8571)^2 + \left(\frac{8}{16}\right)(14025.1429)^2 = 14,912$$

$$t = \frac{(1184 - 1196)}{\sqrt{14912226\left(\frac{1}{8} + \frac{1}{8}\right)}} = \frac{-12}{61.057} = -0.1965$$

Since -0.1965 is not in the rejection region, we fail to reject H_0. There is not sufficient evidence to conclude that there is a difference in the mean tree density for fertilized and control plots.

11.23

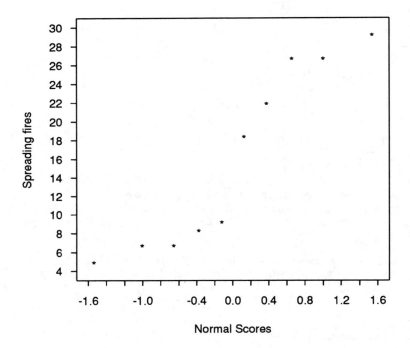

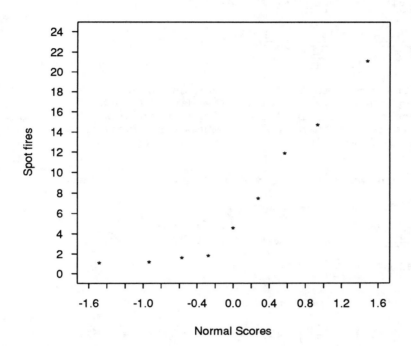

Both graphs show a marked non-linear pattern. These patterns
suggest that neither sample came from a normal population.
However, when n is small, normality should not be ruled out
unless the departure from linearity is very clear cut. The
correlation coefficient between the spreading fires observations
and their normal scores is 0.944 and the correlation coefficient
between the spot fire observations and their normal scores is
0.929. Both of these correlation coefficients are greater than
the critical r for checking for normality. (Refer to chapter 7,
section 3.) Therefore, the departure from linearity in each
graph is not pronounced enough to reject the idea that the
corresponding sample data came from a normal population.
Therefore, the use of the pooled t confidence interval is
justified.

11.25   Let μ_1 denote mean number of imitations for infants who watch a
human model. Let μ_2 denote the mean number of imitations for
infants who watch a doll.

$H_0: \mu_1 - \mu_2 = 0$ $H_a: \mu_1 - \mu_2 > 0$

$$t = \frac{(\overline{x}_1 - \overline{x}_2) - 0}{\sqrt{s_p^2 \left(\frac{1}{n_1} + \frac{1}{n_2} \right)}}$$ with d.f. = 12 + 15 - 2 = 25

147

For $\alpha = 0.01$, reject H_0 if $t > 2.49$.

$$s_p^2 = \left(\frac{11}{25}\right)(1.6)^2 + \left(\frac{14}{25}\right)(1.3)^2 = 2.0728$$

$$t = \frac{(5.14 - 3.46)}{\sqrt{2.0728\left(\frac{1}{12} + \frac{1}{15}\right)}} = \frac{1.68}{.5576} = 3.01$$

Since the computed t of 3.01 exceeds the critical t of 2.49, the null hypothesis is rejected. The data supports the conclusion that the mean number of imitations by infants who watch a human model is larger than the mean number of imitations by infants who watch a doll.

11.27　Let μ_1 denote the true average lung volume for exposed rats and μ_2 the true average lung volume for control rats.

$H_0: \mu_1 - \mu_2 = 0$　　　$H_a: \mu_1 - \mu_2 > 0$

$$t = \frac{(\overline{x}_1 - \overline{x}_2) - 0}{\sqrt{s_p^2\left(\frac{1}{n_1} + \frac{1}{n_2}\right)}}　\text{with d.f.} = 20 + 17 - 2 = 35$$

For $\alpha = 0.01$, reject H_0 if $t > 2.44$.

$$s_p^2 = \left(\frac{19}{35}\right)(.37)^2 + \left(\frac{16}{35}\right)(.41)^2 = .0743 + .0768 = .1512$$

$$t = \frac{(9.28 - 7.97)}{\sqrt{.1512\left(\frac{1}{20} + \frac{1}{17}\right)}} = \frac{1.31}{\sqrt{0.165}} = 10.21$$

Since the calculated t value of 10.21 exceeds the t critical value 2.44, H_0 is rejected. The data strongly supports the conclusion that the mean lung volume of rats exposed to ozone is larger than the mean lung volume of rats not exposed to ozone.

11.29　Let μ_1 and μ_2 denote the mean nitrogen concentration in soil .6m and 6m, respectively, from the roadside.

$H_0: \mu_1 - \mu_2 = 0$　　　$H_a: \mu_1 - \mu_2 > 0$

$$t = \frac{(\overline{x}_1 - \overline{x}_2) - 0}{\sqrt{s_p^2\left(\frac{1}{n_1} + \frac{1}{n_2}\right)}}　\text{with d.f.} = 20 + 20 - 2 = 38$$

For $\alpha = 0.01$, reject H_0 if $t > $ (approx. 2.42).

$$s_p^2 = \frac{(20 - 1)(.4)^2 + (20 - 1)(.3)^2}{20 + 20 - 2} = .125$$

$$t = \frac{(1.70 - 1.35) - 0}{\sqrt{.125\left(\frac{1}{20} + \frac{1}{20}\right)}} = \frac{.35}{.1118} = 3.13$$

Since the computed t of 3.13 exceeds 2.42, H_0 is rejected. The data does provide sufficient evidence to conclude that the mean nitrogen concentration in soil .6m from the roadside is higher than in soil 6m from the roadside.

11.31   Let μ_1 denote the mean number of headache free days using the cognitive-behavioral therapy and μ_2 the mean number of headache free days using the relaxation therapy.

$H_0: \mu_1 - \mu_2 = 0$ $H_a: \mu_1 - \mu_2 > 0$

$$t = \frac{(\overline{x}_1 - \overline{x}_2) - 0}{\sqrt{s_p^2\left(\frac{1}{n_1} + \frac{1}{n_2}\right)}}$$ with d.f. = 24 + 24 - 2 = 46

For α = .05, reject H_0 if t > 1.68 (The value 1.68 is found from the t-table with d.f. = 40.)

$$s_p^2 = \frac{(24 - 1)(1.75)^2 + (24 - 1)(1.43)^2}{24 + 24 - 2} = \frac{117.4702}{46} = 2.5537$$

$$t = \frac{(5.71 - 3.82) - 0}{\sqrt{2.5337\left(\frac{1}{24} + \frac{1}{24}\right)}} = \frac{1.89}{.461} = 4.10$$

Since 4.1 exceeds 1.68, H_0 is rejected. The data does provide sufficient evidence to support the researcher's claim that the cognitive-behavioral therapy is more effective than relaxation in increasing the mean number of headache free days.

11.33   a)   Let μ_1 denote the true mean birth weight for premature infants with brain damage and μ_2 the true mean birth weight for premature infants without brain damage.

$H_0: \mu_1 - \mu_2 = 0$ $H_a: \mu_1 - \mu_2 \neq 0$

$$t = \frac{(\overline{x}_1 - \overline{x}_2) - 0}{\sqrt{s_p^2\left(\frac{1}{n_1} + \frac{1}{n_2}\right)}}$$ with d.f. = 62

For α = 0.05, reject H_0 if t < -2.00 or t > 2.00. (The value of 2.00 was obtained using d.f. = 60.)

$$s_p^2 = \frac{9(448)^2 + 53(656)^2}{62} = 397002.3226$$

$$t = \frac{(1541 - 1204) - 0}{\sqrt{397002.3226\left(\frac{1}{10} + \frac{1}{54}\right)}} = \frac{337}{216.915} = 1.55$$

Since the computed t of = 1.55 does not fall in the rejection region, H_0 is not rejected. There is not sufficient evidence to conclude that there is a difference in the mean birth weight of infants born premature with brain damage and those born premature with no brain damage.

b)   Let μ_1 denote the true mean birth weight for full-term infants with brain damage and μ_2 the true mean birth weight for full-term infants without brain damage.

$H_0: \mu_1 - \mu_2 = 0$ $H_a: \mu_1 - \mu_2 \neq 0$

$$t = \frac{(\overline{x}_1 - \overline{x}_2) - 0}{\sqrt{s_p^2\left(\frac{1}{n_1} + \frac{1}{n_2}\right)}}$$ with d.f. = 12 + 13 - 2 = 23

For α = 0.05, reject H_0 if t < -2.07 or t > 2.07.

$$s_p^2 = \frac{11(707)^2 + 12(627)^2}{23} = 444169$$

$$t = \frac{(2998 - 2704) - 0}{\sqrt{444169\left(\frac{1}{12} + \frac{1}{13}\right)}} = \frac{294}{266.798} = 1.10$$

Since the computed t = 1.1 does not fall in the rejection region, H_0 is not rejected. There is not sufficient evidence to conclude that there is a difference in the mean birth weight of infants carried full term with brain damage and those carried full term without brain damage.

c)   Let μ_1 denote the mean birth weight of infants without brain damage carried full term and μ_2 the mean birth weight of infants without brain damage born premature. The 90% confidence interval for $\mu_1 - \mu_2$ is:

$$(\overline{x}_1 - \overline{x}_2) \pm 1.67 \sqrt{s_p^2\left(\frac{1}{n_1} + \frac{1}{n_2}\right)}$$

where 1.67 is the critical t value associated with 60 d.f.

$$s_p^2 = \frac{(13 - 1)(627)^2 + (54 - 1)(656)^2}{13 + 54 - 2} = 423467.02$$

The confidence interval is:

$(2704 - 1204) \pm 1.67 \sqrt{423467.02\left(\frac{1}{13} + \frac{1}{54}\right)}$

$= 1500 \pm 1.67(201.04) = 1500 \pm 335.73$

$= (1164.27, 1835.73)$.

150

d) Let μ_1 denote the mean birth weight of infants with brain damage carried full term and μ_2 the mean birth weight of infants with brain damage born premature. The 90% confidence interval for $\mu_1 - \mu_2$ is:

$$(\overline{x}_1 - \overline{x}_2) \pm 1.73 \sqrt{s_p^2 \left(\frac{1}{n_1} + \frac{1}{n_2} \right)}$$

$$s_p^2 = \frac{(12 - 1)(707)^2 + (10 - 1)(448)^2}{12 + 10 - 2} = \frac{7304675}{20} = 365233.75$$

The confidence interval is:

$$(2998 - 1541) \pm 1.73 \sqrt{365233.75 \left(\frac{1}{12} + \frac{1}{10} \right)}$$
$$= 1457 \pm 1.73(258.77) = 1457 \pm 447.66 = (1009.34, 1094.66)$$

11.35 a) Let μ_1 be the mean score on student academic development for all unionized schools and μ_2 be the corresponding mean for non-unionized schools.

$H_0: \mu_1 - \mu_2 = 0$ $\qquad$ $H_a: \mu_1 - \mu_2 \neq 0$

$$t = \frac{(\overline{x}_1 - \overline{x}_2) - 0}{\sqrt{s_p^2 \left(\frac{1}{n_1} + \frac{1}{n_2} \right)}} \qquad \text{with d.f.} = 18 + 23 - 2 = 39$$

For $\alpha = 0.05$, reject H_0 if $t > 2.02$ or $t < -2.02$.

$$s_p^2 = \left(\frac{17}{39} \right)(.49)^2 + \left(\frac{22}{39} \right)(.88)^2 = .1047 + .4368 = .5415$$

$$t = \frac{(3.71 - 4.36) - 0}{\sqrt{.5415 \left(\frac{1}{18} + \frac{1}{23} \right)}} = \frac{-0.65}{\sqrt{.0536}} = -2.81$$

Since the calculated t of -2.81 is less than the t critical value -2.02, the null hypothesis is rejected. The data suggests that there is a difference in the mean academic development score of unionized schools and the mean academic development score of non-unionized schools.

b) Let μ_1 be the mean faculty satisfaction score for all unionized schools and μ_2 be the corresponding mean for non-unionized schools.

$H_0: \mu_1 - \mu_2 = 0$ $\qquad$ $H_a: \mu_1 - \mu_2 \neq 0$

$$t = \frac{(\overline{x}_1 - \overline{x}_2) - 0}{\sqrt{s_p^2 \left(\frac{1}{n_1} + \frac{1}{n_2} \right)}} \qquad \text{with d.f.} = 18 + 23 - 2 = 39$$

For α = 0.05, reject H_0 if $t > 2.02$ or $t < -2.02$.

$$s_p^2 = \left(\frac{17}{39}\right)(.56)^2 + \left(\frac{22}{39}\right)(.39)^2 = .1367 + .0858 = .2225$$

$$t = \frac{(4.49 - 4.85) - 0}{\sqrt{.2225\left(\frac{1}{18} + \frac{1}{23}\right)}} = \frac{-0.36}{\sqrt{.022}} = -2.43$$

Since the calculated t of -2.43 is less than the t critical value -2.02, the null hypothesis is rejected. The data suggests that there is a difference between the mean satisfaction score of faculty at unionized schools and the mean satisfaction score of faculty at non-unionized schools.

c) The absolute value of the calculated t is 2.43, which is between 2.42 and 2.70 in the t table. For a two-tailed test, these t values have α values of .02 and .01 respectively. Therefore, it can be said that $0.02 > $ P-value $> .01$.

11.37 a) Two conditions are required to appropriately apply the pooled t-test.

      i) Both population distributions are normal.
     ii) The two population variances are the same.

A check of the summary statistics can help in assuring whether either of these two conditions seems to be violated. Looking at the data, $s_1 = .94$ and $s_2 = .98$. Without using a formal procedure, it certainly appears plausible that the population standard deviations σ_1 and σ_2 are equal. That is, there is insufficient evidence to suggest otherwise. Without the original data, it is more difficult to check on condition (i). However, there appears to be nothing in the summary statistics to suggest that (i) does not hold. Hence, the use of the pooled t-test does seem reasonable in this investigation.

b) Let $\mu_1 - \mu_2$ denote the mean oxygen consumption for the untreated and treated beetles, respectively.

$H_0: \mu_1 - \mu_2 = 0$ $H_a: \mu_1 - \mu_2 > 0$

$$t = \frac{(\overline{x}_1 - \overline{x}_2) - 0}{\sqrt{s_p^2\left(\frac{1}{n_1} + \frac{1}{n_2}\right)}} \qquad \text{with d.f.} = 25 + 25 - 2 = 48$$

For α = 0.05, reject H_0 if $t > 1.68$ (approx.).

$$s_p^2 = \left(\frac{24}{48}\right)(.94)^2 + \left(\frac{24}{48}\right)(.98)^2 = \frac{1}{2}[(.94)^2 + (.98)^2] = .922$$

$$t = \frac{(5.02 - 4.37) - 0}{\sqrt{.922\left(\frac{1}{25} + \frac{1}{25}\right)}} = \frac{.65}{\sqrt{0.07376}} = 2.39$$

Since the t calculated of 2.39 exceeds the t critical 1.68, H_0 is rejected. There is sufficient evidence to conclude that the mean oxygen consumption is less when the DDT treatment is used.

c)   From Appendix Table IV, .025 > P-value > .01. Since the P-value exceeds .01, the null hypothesis would not be rejected at the .01 level of significance.

11.39   a)   No. the z test of Section 2 should used only if both n_1 and n_2 are at least 30.

b)   Let μ_1 and μ_2 denote the true mean cadmium skull concentrations for the control group and the reclamation group, respectively.

$$H_0: \mu_1 - \mu_2 = 0 \qquad H_a: \mu_1 - \mu_2 \neq 0$$

$$t = \frac{(\overline{x}_1 - \overline{x}_2) - 0}{\sqrt{s_p^2\left(\frac{1}{n_1} + \frac{1}{n_2}\right)}} \qquad \text{with d.f.} = 20 + 21 - 2 = 39$$

For $\alpha = 0.05$, reject H_0 if t > 2.02 or t < -2.02.

$$s_p^2 = \left(\frac{19}{39}\right)(.13)^2 + \left(\frac{20}{39}\right)(.18)^2 = .0248$$

$$t = \frac{(.59 - .72) - 0}{\sqrt{.0248\left(\frac{1}{20} + \frac{1}{21}\right)}} = \frac{-.13}{.0492} = -2.64$$

Since the calculated of -2.64 is less than the t critical -2.02, H_0 is rejected. The data suggests that the true mean cadmium skull concentrations differ for the two sites.

11.41   a) $s_p^2 = \left[\frac{6 - 1}{6 + 12 - 2}\right](78.4)^2 + \left[\frac{12 - 1}{6 + 12 - 2}\right](65.8)^2$

$$= \left(\frac{5}{16}\right)(78.4)^2 + \left(\frac{11}{16}\right)(65.8)^2 = 4897.4275$$

b)   Let μ_1 and μ_2 denote the true average yield when not under water stress and under water stress, respectively.

$$H_0: \mu_1 - \mu_2 = 0 \qquad H_a: \mu_1 - \mu_2 > 0$$

$$t = \frac{(\overline{x}_1 - \overline{x}_2) - 0}{\sqrt{s_p^2\left(\frac{1}{n_1} + \frac{1}{n_2}\right)}} \qquad \text{with d.f.} = 6 + 12 - 2 = 16$$

For $\alpha = 0.01$, reject H_0 if $t > 2.58$.

$$t = \frac{(376 - 234) - 0}{\sqrt{4897.4275\left(\frac{1}{6} + \frac{1}{12}\right)}} = \frac{142}{34.99} = 4.058$$

Since the calculated t of 4.058 exceeds the t critical 2.58, H_0 is rejected. The data does suggest that the true average yield under stress is less than that for the unstressed condition.

c)   The P-value is less than .0005.

11.43   a)   Let π_1 and π_2 denote the proportions of all male and female shoppers, respectively, who buy only name-brand grocery products.

$H_0: \pi_1 - \pi_2 = 0 \qquad H_a: \pi_1 - \pi_2 \neq 0$

$$z = \frac{p_1 - p_2}{\sqrt{\dfrac{p_c(1-p_c)}{n_1} + \dfrac{p_c(1-p_c)}{n_2}}}$$

For $\alpha = 0.05$, reject H_0 if $z > 1.96$ or if $z < -1.96$.

$$p_1 = \frac{87}{200} = .435 \qquad p_2 = \frac{96}{300} = .320$$

$$p_c = \left(\frac{200}{500}\right)(.435) + \left(\frac{300}{500}\right)(.320) = .366$$

$$z = \frac{(.435 - .320)}{\sqrt{\dfrac{.366(.634)}{200} + \dfrac{.366(.634)}{300}}} = \frac{.115}{.044} = 2.62$$

Since the z calculated value of 2.62 exceeds the z critical 1.96, H_0 is rejected. The data suggests that there is a difference between the proportion of males who buy only name-brand grocery products and that of females.

b) $(.435 - .32) \pm 1.96\sqrt{\dfrac{.435(.565)}{200} + \dfrac{.320(.680)}{300}}$

$(.115) \pm 1.96(.044) = .115 \pm .087 = (.028, .202)$

With 95% confidence, it is estimated that the proportion of males who buy only name-brand grocery products exceeds that of females by as little as 2.8% or possibly by as much as 20.2%.

11.45   Let π_1 denote the proportion of returned surveys when a plain cover is used and π_2 denote the proportion of all returned surveys when a picture of a skydiver is used on the cover.

$H_0: \pi_1 - \pi_2 = 0 \qquad H_a: \pi_1 - \pi_2 < 0$

$$z = \frac{p_1 - p_2}{\sqrt{\dfrac{p_c(1-p_c)}{n_1} + \dfrac{p_c(1-p_c)}{n_2}}}$$

For $\alpha = .10$, reject H_0 if $z < -1.28$.

$$p_1 = \frac{104}{207} = .5024 \quad \text{and} \quad p_2 = \frac{109}{213} = .5117$$

$$p_c = \left(\frac{207}{420}\right)(.5024) + \left(\frac{213}{420}\right)(.5117) = .5071$$

$$z = \frac{(.5024 - .5117)}{\sqrt{\frac{.5071(.4929)}{207} + \frac{.5071(.4929)}{213}}} = \frac{-.0093}{.0448} = -.1906$$

Since -0.1906 is not less than -1.28, the null hypothesis is not rejected. The data do not support the researcher's claim that the response rate for the plain cover survey is lower than the response rate for the survey whose cover carried a picture of a skydiver.

11.47   Let π_1 denote the proportion of women in 1970 who had never married and π_2 denote the proportion of women in 1990 who had never married.

$H_0: \pi_1 - \pi_2 = 0$ $\qquad$ $H_a: \pi_1 - \pi_2 < 0$

$$z = \frac{p_1 - p_2}{\sqrt{p_c(1-p_c)\left[\frac{1}{n_1} + \frac{1}{n_2}\right]}}$$

For $\alpha = 0.10$, reject H_0 if $z < -1.28$.

$p_1 = .06$, $P_2 = .16$ and $p_c = \frac{(12 + 32)}{200 + 200} = .11$

$$z = \frac{(.06 - .16)}{\sqrt{.11(.89)\left[\frac{1}{200} + \frac{1}{200}\right]}} = \frac{-.10}{.0313} = -3.195$$

Since -3.195 is less than -1.28, the null hypothesis is rejected. This supports the conclusion that the proportion of women in 1990 who had never married exceeds the proportion of women in 1970 who had never married.

11.49   Let π_1 denote the true proportion of males who pump their own gas. Let π_2 denote the true proportion of females who pump their own gas.

$$(p_1 - p_2) \pm (z \text{ critical}) \sqrt{\frac{p_1(1-p_1)}{n_1} + \frac{p_2(1-p_2)}{n_2}}$$

$p_1 = .85$, $p_2 = .70$

The 90% confidence interval for $\pi_1 - \pi_2$ is:

$$(.85 - .70) \pm (1.645) \sqrt{\frac{(.85)(.15)}{100} + \frac{(.70)(.30)}{100}}$$
$$(.15) \pm (1.645)(.0581) = .15 \pm .0956 = (.0544, .2456)$$

11.51   Let π_1 denote the proportion of games where a player suffers a sliding injury when stationary bases are used. Let π_2 denote the proportion of games where a player suffers a sliding injury when break-away bases are used.

$H_0: \pi_1 - \pi_2 = 0 \qquad H_a: \pi_1 - \pi_2 > 0$

$$z = \frac{p_1 - p_2}{\sqrt{\dfrac{p_c(1-p_c)}{n_1} + \dfrac{p_c(1-p_c)}{n_2}}}$$

For $\alpha = 0.01$, reject H_0 if $z > 2.33$.

$p_1 = \dfrac{90}{1250} = .072, \; P_2 = \dfrac{20}{1250} = .016,$

$p_c = \dfrac{90 + 20}{1250 + 1250} = \dfrac{110}{2500} = .044$

$$z = \frac{(.072 - .016)}{\sqrt{\dfrac{.044(.956)}{1250} + \dfrac{.044(.956)}{1250}}} = \frac{.056}{.0082} = 6.83$$

Since the calculated value of 6.83 exceeds the z critical of 2.33, H_0 is rejected. The data suggests that the use of break-away bases reduces the proportion of games in which a player suffers a sliding injury.

11.53   Let π_1 denote the proportion of irradiated garlic bulbs that are marketable and π_2 denote the proportion of untreated bulbs that are marketable.

$H_0: \pi_1 - \pi_2 = 0 \qquad H_a: \pi_1 - \pi_2 > 0$

$$z = \frac{p_1 - p_2}{\sqrt{\dfrac{p_c(1-p_c)}{n_1} + \dfrac{p_c(1-p_c)}{n_2}}}$$

For $\alpha = 0.01$, reject H_0 if $z > 2.33$.

$p_1 = \dfrac{153}{180} = .85, \; p_2 = \dfrac{119}{180} = .6611, \; p_c = .7556$

$$z = \frac{(.85 - .6611)}{\sqrt{\dfrac{.7556(.2444)}{180} + \dfrac{.7566(.2444)}{180}}} + \frac{.1889}{.0453} = 4.17$$

Since the z calculated of 4.17 exceeds the z critical value 2.33, H_0 is rejected. Yes, the data suggests quite strongly that the true proportion of marketable irradiated bulbs exceeds that for untreated bulbs.

11.55   Let π_1 denote the true proportion of qualifying divers who exercise within one hour of competition. Let π_2 denote the true proportion of nonqualifying divers who exercise within one hour of competition.

$H_0: \pi_1 - \pi_2 = 0 \qquad H_a: \pi_1 - \pi_2 \neq 0$

$$z = \frac{p_1 - p_2}{\sqrt{\frac{p_c(1-p_c)}{n_1} + \frac{p_c(1-p_c)}{n_2}}}$$

For $\alpha = .10$, reject H_0 if $z > 1.645$ or if $z < -1.645$.

$p_1 = \frac{7}{20} = .35$, $p_2 = \frac{12}{25} = .48$,

$p_c = \left(\frac{20}{45}\right)(.35) + \left(\frac{25}{45}\right)(.48) = .4222$

$$z = \frac{(.35 - .48)}{\sqrt{\frac{.4222(.5778)}{20} + \frac{.4222(.5778)}{25}}} = \frac{-.13}{.148} = -.88$$

Since the z calculated of $-.88$ falls between the two z critical values, H_0 is not rejected. There is insufficient evidence to indicate that the true proportion of qualifying divers who exercise within one hour of the competition differs from the corresponding proportion for nonqualifying divers.

11.57 Let π_1 denote the proportion of babies to fall victim of SIDS whose mothers use cocaine. Let π_2 denote the proportion of babies to fall victim of SIDS whose mothers do not use cocaine.

$H_0: \pi_1 - \pi_2 = 0$ $H_a: \pi_1 - \pi_2 > 0$

$$z = \frac{p_1 - p_2}{\sqrt{\frac{p_c(1-p_c)}{n_1} + \frac{p_c(1-p_c)}{n_2}}}$$

For $\alpha = 0.01$, reject H_0 if $z > 2.33$.

$p_1 = \frac{10}{60} = .1667$, $p_2 = \frac{5}{1600} = .0031$,

$p_c = \frac{10 + 5}{60 + 1600} = .00904$

$$z = \frac{(.1667 - .0031)}{\sqrt{\frac{.00904(.99096)}{60} + \frac{.00904(.99096)}{1600}}} = \frac{.1636}{.0124} = 13.19$$

Since 13.19 exceeds 2.33, H_0 is rejected. The data strongly supports the conclusion that the proportion of babies to fall victim to SIDS whose mothers use cocaine is larger than that for babies whose mothers do not use cocaine.

11.59 a) Let π_1 denote the true proportion of Californians that favored a state lottery in 1983. Let π_2 denote the true proportion of Californians that favored a state lottery in 1971.

$H_0: \pi_1 - \pi_2 = 0 \qquad H_a: \pi_1 - \pi_2 > 0$

$$z = \frac{p_1 - p_2}{\sqrt{\dfrac{p_c(1-p_c)}{n_1} + \dfrac{p_c(1-p_c)}{n_2}}}$$

For $\alpha = 0.01$, reject H_0 if $z > 2.33$.

$p_1 = \dfrac{578}{750} = .7707$, $p_2 = \dfrac{518}{750} = .6907$,

$p_c = \left(\dfrac{750}{1500}\right)(.7707) + \left(\dfrac{750}{1500}\right)(.6907) = .7307$

$$z = \frac{(.7707 - .6907)}{\sqrt{\dfrac{.7307(.2693)}{750} + \dfrac{.7307(.2693)}{750}}} = \frac{.08}{.0229} = 3.49$$

Since the z calculated of 3.49 exceeds the z critical value 2.33, H_0 is rejected. The data does strongly suggest that the true proportion of Californians that favored a state lottery in 1983 exceeds the corresponding 1971 proportion.

b)   P-value = $P(z > 3.49) = .0002$

11.61   Let μ_1 denote the true average fluoride concentration for livestock grazing in the polluted region and μ_2 denote the true average fluoride concentration for livestock grazing in the unpolluted regions.

$H_0: \mu_1 - \mu_2 = 0$ $H_a: \mu_1 - \mu_2 > 0$

The test statistic is: rank sum for polluted area.

For $\alpha = 0.05$, reject H_0 if rank sum ≥ 56.

Sample	Ordered Data	Rank
2	14.2	1
1	16.8	2
1	17.1	3
2	17.2	4
2	18.3	5
2	18.4	6
1	18.7	7
1	19.7	8
2	20.0	9
1	20.9	10
1	21.3	11
1	23.0	12

Rank sum $= (2 + 3 + 7 + 8 + 10 + 11 + 12) = 53$

Since the rank sum of 53 does not fall into the rejection region, H_0 is not rejected. The data does not support the conclusion that there is a larger average fluoride concentration for the polluted area than for the unpolluted area.

11.63   a)   Let μ_1 denote the true average ascent time using the lateral gait and μ_2 denote the true average ascent time using the four-beat diagonal gait.

$H_0: \mu_1 - \mu_2 = 0$ $H_a: \mu_1 - \mu_2 \neq 0$

The test statistic is: Rank sum for diagonal gait.

For $\alpha = 0.05$,
    reject H_0 if rank sum ≤ 28 or if rank sum ≥ 56.

Gait	Ordered Data	Rank
D	.85	1
L	.86	2
L	1.09	3
D	1.24	4
D	1.27	5
L	1.31	6
L	1.39	7
D	1.45	8
L	1.51	9
L	1.53	10
L	1.64	11
D	1.66	12
D	1.82	13

Rank sum = 1 + 4 + 5 + 8 + 12 + 13 ≐ 43

Since the rank sum value of 43 does not fall into the rejection region, H_0 is not rejected. The data does not suggest that there is a difference in mean ascent time for the diagonal and lateral gaits.

b)   From Appendix Table VII, d = 8 is found n_1 = 6, n_2 = 7 and confidence level of 95%.

Differences

Lateral gait

		.86	1.09	1.31	1.39	1.51	1.53	1.64
	.85	.01	.24	.46	.54	.66	.68	.79
	1.24	−.38	−.15	.07	.15	.27	.29	.40
Diagonal	1.27	−.41	−.18	.04	.12	.24	.26	.37
gait	1.45	−.59	−.36	−.14	−.06	.06	.08	.19
	1.66	−.80	−.57	−.35	−.27	−.15	−.13	−.02
	1.82	−.96	−.73	−.51	−.43	−.31	−.29	−.18

The confidence interval for $\mu_1 - \mu_2$ is (−.41, .29).

11.65   Let μ_1 denote the true mean number of errors for students who plagiarize and μ_2 the true mean number of errors for students who do not plagiarize.

H_0: $\mu_1 - \mu_2 = 0$ H_a: $\mu_1 - \mu_2 > 0$

The test statistic is: $z = \dfrac{\text{rank sum} - \dfrac{n_1(n_1+n_2+1)}{2}}{\sqrt{\dfrac{n_1 n_2 (n_1+n_2+1)}{12}}}$.

For α = 0.01, reject H_0 if z > 2.33.

n_1 = 39 (plagiarism), n_2 = 35 (no plagiarism)

$$z = \frac{(1795.5) - \dfrac{39(39+35+1)}{2}}{\sqrt{\dfrac{39(35)(39+35+1)}{12}}} = \frac{1795.5 - 1462.5}{\sqrt{8531.25}} = \frac{333}{92.36} = 3.61$$

Since the z calculated value of 3.61 falls in the rejection region, H_0 is rejected. The data suggests that students who do their own writing are better able to reproduce their writing than those who had copied someone else's writing.

11.67 Let μ_1 denote the true mean number of binges per week for people who use Imipramine and μ_2 the true mean number of binges per week for people who use a placebo.

H_0: $\mu_1 - \mu_2 = 0$ $\qquad$ H_a: $\mu_1 - \mu_2 < 0$

The test statistic is: Rank sum for the Imipramine group.

For $\alpha = 0.05$, reject H_0 if rank sum ≤ 52.

Gait	Ordered Data	Rank
I	1	1.5
I	1	1.5
I	2	3.5
I	2	3.5
I	3	6
P	3	6
P	3	6
P	4	8.5
P	4	8.5
I	5	10
P	6	11
I	7	12
P	8	13
P	10	14
I	12	15
P	15	16

Rank sum = 1.5 + 1.5 + 3.5 + 3.5 + 6 + 10 + 12 + 15 = 53

Since the rank sum of 53 does not fall into the rejection region, H_0 is not rejected. The data suggests that Imipramine is not effective in reducing the mean number of binges per week.

11.69 Let μ_1 denote the true mean length of odontablasts in guinea pigs who receive orange juice and μ_2 the corresponding mean for guinea pigs who receive Ascorbic acid.

H_0: $\mu_1 - \mu_2 = 0$ $\qquad$ H_a: $\mu_1 - \mu_2 \neq 0$

$$z = \frac{\text{rank sum} - \dfrac{n_1(n_1+n_2+1)}{2}}{\sqrt{\dfrac{n_1 n_2(n_1+n_2+1)}{12}}}$$

For $\alpha = 0.01$, reject H_0 if $z > 2.58$ or if $z < -2.58$.

Rank sum for orange juice group = 125

$$z = \frac{125 - \dfrac{10(20)}{2}}{\sqrt{\dfrac{10(9)(20)}{12}}} = \frac{125 - 1400}{\sqrt{150}} = \frac{25}{12.25} = 2.04$$

Since the z calculated value of 2.04 does not fall into the rejection region, H_0 is not rejected. The data suggests that there is not a difference in the mean length of odontablasts for the two types of vitamin C intake.

11.71   Let μ_1 denote the mean burn time for oak and μ_2 the mean burn time for pine. From Appendix Table VII d = 9,

Differences

Oak

		.48	.67	1.23	1.42	1.55	1.56	1.72	1.77
	.73	- 2.5	-.06	.50	.69	.82	.83	.99	1.04
	.98	- .50	-.31	.25	.44	.57	.58	.74	.79
Pine	1.20	- .72	-.53	.03	.22	.35	.36	.52	.57
	1.33	- .85	-.66	-.10	.09	.22	.23	.39	.44
	1.40	- .92	-.73	-.17	.02	.15	.16	.32	.37
	1.52	-1.04	-.85	-.29	-.10	.03	.04	.20	.25

The approximate 95% distribution-free confidence interval on the difference of the mean burning times of oak and pine is from -.5 to 0.57. The confidence interval indicates that the mean burning time of oak may be as much as .57 hours longer than pine, but also that the mean burning time of oak may be as much as .5 hours shorter than pine.

11.73 Let μ_1 and μ_2 denote the mean pH level of the soil at locations A and B, respectively.

$H_0: \mu_1 - \mu_2 = 0$ $H_a: \mu_1 - \mu_2 \neq 0$

The test statistic is:

$$t = \frac{(\bar{x}_1 - \bar{x}_2) - 0}{\sqrt{s_p^2\left(\frac{1}{n_1} + \frac{1}{n_2}\right)}}$$ with d.f. = 9 + 9 − 2 = 16.

For $\alpha = 0.05$, reject H_0 if $t < -2.12$ or if $t > 2.12$.

For location A: $\bar{x}_1 = 8.0378$, $s_1^2 = .0813$
For location B: $\bar{x}_2 = 7.4422$, $s_2^2 = .0504$

$$s_p^2 = \left(\frac{8}{16}\right)(.0813) + \left(\frac{8}{16}\right)(.0504) = .06585$$

$$t = \frac{(8.0378 - 7.4422)}{\sqrt{.06585\left(\frac{1}{9} + \frac{1}{9}\right)}} = \frac{.5956}{.12097} = 4.92$$

Since the calculated t value of 4.92 falls in the rejection region, the null hypothesis is rejected. The data does suggest that the mean soil pH levels differ for the two locations. The conditions necessary for this test to be valid are:

1. The distribution of pH must be normal at each of the locations.

2. The standard deviations of these two distributions must be equal.

11.75 Let μ_1 denote the mean score of male faculty on the MSQ and μ_2 the mean score of female faculty on the MSQ.

$H_0: \mu_1 - \mu_2 = 0$ $H_a: \mu_1 - \mu_2 \neq 0$

The test statistic is: $z = \dfrac{(\bar{x}_1 - \bar{x}_2) - 0}{\sqrt{\dfrac{s_1^2}{n_1} + \dfrac{s_2^2}{n_2}}}$.

For $\alpha = 0.01$, reject H_0 if $z < -2.58$ or if $z > 2.58$.

$$z = \frac{(75.43 - 72.54)}{\sqrt{\dfrac{(10.53)^2}{115} + \dfrac{(13.08)^2}{105}}} = 1.79$$

Since the z calculated value of 1.79 does not fall in the rejection region, the null hypothesis is not rejected. The data does not suggest that male and female academic employees differ with respect to mean score on the MSQ.

164

11.77 Let μ_1 denote the true mean age at death for female SIDS victims and μ_2 the true mean age at death for male SIDS victims.

From the data,
$\overline{x} = 103.6$, $s_1^2 = 2154.3$, $\overline{x}_2 = 89.7$, $s_2^2 = 1638.238$
$$s_p^2 = \left(\frac{4}{10}\right)(2154.3) + \left(\frac{6}{10}\right)(1638.238) = 1844.66$$

The confidence interval is:
$$(103.6 - 89.7) \pm (2.23)\sqrt{\frac{1844.66}{5} + \frac{1844.66}{7}}$$
$(13.9) \pm (2.23)(25.149) = 13.9 \pm 56.08 = (-42.18, \ 69.98)$

With 95% confidence, it is estimated that $\mu_1 - \mu_2$ is between -42.18 and 69.98. Since this interval contains zero, the analysis supports the conclusion that there may be no difference between the true mean ages at death of male and female SIDS victims.

11.79 For each of the tests, let μ_1 denote the mean blood characteristic for the mice exposed to an electric field and μ_2 the mean blood characteristic for the control group.

H_0: $\mu_1 - \mu_2 = 0$ $\qquad$ H_a: $\mu_1 - \mu_2 \neq 0$

The test statistic is: $z = \dfrac{(\overline{x}_1 - \overline{x}_2) - 0}{\sqrt{\dfrac{s_1^2}{n_1} + \dfrac{s_2^2}{n_2}}}$.

For $\alpha = 0.05$, reject H_0 if $z < -1.96$ or if $z > 1.96$.

For glucose:

$$z = \frac{(139.20 - 136.30)}{\sqrt{\dfrac{(16.1)^2}{45} + \dfrac{(12.7)^2}{45}}} = \frac{2.90}{3.06} = .95$$

Since the z calculated value of .95 does not fall in the rejection region, the null hypothesis is not rejected. The data suggests that there is no difference in mean glucose levels of the two populations.

For potassium:

$$z = \frac{(7.62 - 7.44)}{\sqrt{\dfrac{(.54)^2}{45} + \dfrac{(.53)^2}{45}}} = \frac{0.180}{0.113} = 1.60$$

Since the z calculated value of 1.6 does not fall in the rejection region, the null hypothesis is not rejected. The data suggests that there is no difference in mean potassium levels for the two populations.

For total protein:

$$z = \frac{(6.61 - 6.63)}{\sqrt{\dfrac{(.34)^2}{45} + \dfrac{(.27)^2}{45}}} = \frac{-.02}{.065} = -.31$$

Since the z calculated value of -.31 does not fall in the rejection region, the null hypothesis is not rejected. The data suggests that there is no difference in mean total protein levels for the two populations.

For cholesterol:

$$z = \frac{(67.8 - 69.0)}{\sqrt{\dfrac{(9.38)^2}{45} + \dfrac{(11.4)^2}{45}}} = \frac{-1.2}{2.2} = -0.55$$

Since the z calculated value of -0.55 does not fall in the rejection region, the null hypothesis is not rejected. The data suggests that there is no difference in mean cholesterol levels for the two populations.

11.81   a)   Let μ_1 denote the true mean change in SEI score for inmates who receive mathematics tutoring. Let μ_2 denote the true mean change in SEI score for inmates who do not receive mathematics tutoring.

$H_0: \mu_1 - \mu_2 = 0$ $H_a: \mu_1 - \mu_2 > 0$

The test statistic is: $z = \dfrac{(\overline{x}_1 - \overline{x}_2) - 0}{\sqrt{\dfrac{s_1^2}{n_1} + \dfrac{s_2^2}{n_2}}}$.

For $\alpha = 0.01$, reject H_0 if $z > 2.33$.

$$z = \frac{[2.9 - (-1.3)]}{\sqrt{\dfrac{(5.4)^2}{40} + \dfrac{(5.6)^2}{40}}} = \frac{4.2}{1.23} = 3.41$$

Since the z calculated value of 3.41 falls in the rejection region, the null hypothesis is rejected. The data does provide sufficient evidence to conclude that the mean change in SEI scores is larger for inmates who receive mathematics tutoring than it is for inmates who do not receive mathematics tutoring.

    b)   P-value = $P(z > 3.41) = .0003$

11.83   Let π_1 denote the true proportion of adults born deaf who remove the implants. Let π_2 denote the true proportion of adults who went deaf after learning to speak who remove the implants.

$H_0: \pi_1 - \pi_2 = 0$ $H_a: \pi_1 - \pi_2 \neq 0$

The test statistic is: $z = \dfrac{(p_1 - p_2)}{\sqrt{\dfrac{p_c(1-p_c)}{n_1} + \dfrac{p_c(1-p_c)}{n_2}}}$.

For $\alpha = 0.01$, reject H_0 if $z < -2.58$ or if $z > 2.58$.

From the data, $p_1 = \dfrac{75}{250} = .3$, $p_2 = \dfrac{25}{250} = .1$,

$p_c = \left(\dfrac{250}{500}\right)(.3) + \left(\dfrac{250}{500}\right)(.1) = .2$

$z = \dfrac{.3 - .1}{\sqrt{\dfrac{(.28)(.8)}{250} + \dfrac{(.2)(.8)}{250}}} = \dfrac{.2}{.03577} = 5.59$

Since the calculated z value of 5.59 falls in the rejection region, the null hypothesis is rejected. The data does support the fact that the true proportion who remove the implants differs from those that were born deaf from that of those who went deaf after learning to speak.

CHAPTER 12
INFERENCES USING PAIRED DATA

Section 12.1

12.1   The difference between independent samples and paired data is
        that in paired data an observation in the first sample is
        associated with or related to an observation in the second
        sample, whereas, for independent samples each observation in the
        first sample is not related or dependent on any of the
        observations in the second sample.

12.3   To determine if the training regimen is successful in improving
        performance, the director of the sports camp could proceed in
        the following manner. Randomly select some of the participants
        who are training for the 100 meter freestyle. Measure each
        selected participant's time before being placed on the camp's
        training program. Then, after a specified length of time on the
        training program, measure each selected participant's time.
        Thus, there are before and after times for each selected
        participant. This would result in paired samples.

12.5   a)   If possible, treat each patient with both drugs. One drug
              used on one eye, the other drug used on the other eye.
              Then take observations (readings) of eye pressure on each
              eye. If this treatment method is not possible, then
              request the opthamologist to pair patients according to
              their eye pressure so that the two people in a pair have
              approximately equal eye pressure. Then treat one of the
              patients in the pair with the new drug and record the
              reduction in eye pressure. Treat the other person in that
              pair with the standard treatment and record the reduction
              in eye pressure. These two readings would constitute a
              pair. Repeat for each of the other pairs to obtain the
              paired sample data.

        b)   Both procedures above would result in paired data.

        c)   Select a group of persons to participate in the study.
              Randomly select a subset of this group to receive the new
              drug, and give the standard treatment to the remaining
              people. Measure reduction in eye pressure for both groups.
              The resulting observations would constitute independent
              samples.

12.7   For a paired data experiment:

        Ten runners are chosen to participate in the project. Each
        runner is given two pairs of shoes, one pair of brand A and one
        pair of brand B. For each runner, a brand is randomly selected
        to be worn on the right foot. Thus, each runner is to be wearing
        one shoe of each brand. Two observations are made on each runner
        and thus paired data results.

168

For an independent samples experiment:

Twenty runners are chosen to participate in the project. Ten runners are randomly chosen to use brand A, with the remaining ten runners using brand B. Thus, each runner wears the same brand on both feet. One reading is taken for each runner. The resulting two samples of size ten are each independent samples.

The paired data experiment would be recommended because the large variation in shoe wear from runner to runner would tend to obscure any difference between brands.

12.9　　Let μ_d denote the true average difference in yield between the two varieties of wheat. (Sundance - Manitou)

　　　　H_0: $\mu_d = 0$ (no difference in average yield)
　　　　H_a: $\mu_d > 0$ (average yield for Sundance is larger than average yield for Manitou)

　　　　The test statistic is:

$$t = \frac{\overline{x}_d - 0}{\frac{s_d}{\sqrt{n}}} \quad \text{with d.f.} = 8$$

　　　　For $\alpha = .01$, reject H_0 if $t > 2.90$.

　　　　The differences are: 815, 1084, 681, 550, 535, 786, 1162, 517, 910.

　　　　From these: $\overline{x}_d = 782.222$ and $s_d = 236.736$

$$t = \frac{782.222 - 0}{\frac{236.736}{\sqrt{9}}} = 9.91$$

　　　　Since the t calculated of 9.91 exceeds the t critical value 2.90, the null hypothesis is rejected. The data supports the conclusion that the mean yield of Sundance exceeds the mean yield of Manitou.

12.11　　Let μ_d denote the true average difference in number of seeds detected by the two methods. (Direct - Stratified)

　　　　H_0: $\mu_d = 0$ (no difference in average number of seeds detected)
　　　　H_a: $\mu_d \neq 0$ (average number of seeds detected by the Direct method is not the same as the average number of seeds detected by the Stratified method)

　　　　The test statistic is:

$$t = \frac{\overline{x}_d - 0}{\frac{s_d}{\sqrt{n}}} \quad \text{with d.f.} = 26$$

　　　　For $\alpha = .05$, reject H_0 if $t < -2.06$ or if $t > 2.06$.

　　　　The differences are: 16, -4, -8, 4, -32, 0, 12, 0, 4, -8, 4, 12, 8, -28, 4, 0, 0, 4, 0, -8, -8, 0, 0, -4, -28, 4, -36.

　　　　From these: $\overline{x}_d = -3.407$ and $s_d = 13.253$

$$t = \frac{-3.407 - 0}{\frac{13.253}{\sqrt{27}}} = -1.34$$

Since the t calculated of -1.34 is between -2.06 and 2.06, the null hypothesis is not rejected. The data do not provide sufficient evidence to conclude that the mean number of seeds detected differs for the two methods.

12.13   Let μ_d denote the difference between the true mean time of useful consciousness when no alcohol is consumed and that when .5 cc of alcohol per pound of body weight is ingested. (No alcohol - Alcohol)

$H_0: \mu_d = 0$ $H_a: \mu_d > 0$

The test statistic is:

$t = \dfrac{\overline{x}_d - 0}{\dfrac{s_d}{\sqrt{n}}}$ with d.f. = 9

For α = .05, reject H_0 if t > 1.83.

From the differences: $\overline{x}_d = 195.6$ and $s_d = 230.53$

$t = \dfrac{195.6 - 0}{\dfrac{230.53}{\sqrt{10}}} = 2.68$

Since the calculated t of 2.68 exceeds the t critical of 1.83, the null hypothesis is rejected at level .05. There is sufficient evidence to indicate that ingestion of .5 cc of whiskey per pound of body weight reduces the average time of useful consciousness.

12.15   Let μ_d denote the true average difference in time to parallel park for designs 1 and 2. (Design 1 - Design 2)

$H_0: \mu_d = 0$ $H_a: \mu_d \neq 0$

The test statistic is:

$t = \dfrac{\overline{x}_d - 0}{\dfrac{s_d}{\sqrt{n}}}$ with d.f. = 13

For α = .10, reject H_0 if t < -1.77 or t > 1.77

The differences are: 19.2, 5.6, -.6, -17.2, .6, -5, 7, 26, 5.8, 1.2, -6.2, .6, 4, -24.

From these: $\overline{x}_d = 1.2143$ and $s_d = 12.685$

$t = \dfrac{1.2143 - 0}{\dfrac{12.685}{\sqrt{14}}} = .36$

The computed t of .36 does not fall in the rejection region so H_0 is not rejected. There is not sufficient evidence to conclude that the mean time to parallel park differs significantly for the two designs.

12.17   a)   Let μ_d denote the mean difference in blood pressure. (dental setting minus medical setting)

$H_0: \mu_d = 0$ $H_a: \mu_d > 0$

The test statistic is:

$$t = \frac{\overline{x}_d - 0}{\frac{s_d}{\sqrt{n}}} \quad \text{with d.f.} = 59$$

For $\alpha = .01$, reject H_0 if $t > 2.39$.

$$t = \frac{4.47 - 0}{\frac{8.77}{\sqrt{60}}} = 3.95$$

Since the calculated t of 3.95 exceeds the t critical value 2.39, H_0 is rejected. Thus, the data does suggest that true mean blood pressure is higher in a dental setting than in a medical setting.

b)   Let μ_d denote the true mean difference in pulse rate. (dental minus medical)

$H_0: \mu_d = 0$ $H_a: \mu_d \neq 0$

The test statistic is:

$$t = \frac{\overline{x}_d - 0}{\frac{s_d}{\sqrt{n}}} \quad \text{with d.f.} = 59$$

For $\alpha = .05$, reject H_0 if $t < -2.00$ or if $t > 2.00$.

$$t = \frac{-1.33 - 0}{\frac{8.84}{\sqrt{60}}} = -1.17$$

Since the t calculated value does not fall in the rejection region, H_0 is not rejected. There is not sufficient evidence to conclude that mean pulse rates differ for a dental setting and a medical setting.

12.19   Let μ_d denote the mean difference in number of words recalled. (one-hour minus twenty-four hours)

H_0: $\mu_d = 3$ H_a: $\mu_d > 3$

The test statistic is:

$$t = \frac{\overline{x}_d - 3}{\dfrac{s_d}{\sqrt{n}}} \quad \text{with d.f.} = 7$$

For $\alpha = .01$, reject H_0 if $t > 3.00$.

The differences are: 4, 8, 4, 1, 2, 3, 4, 3.

From these: $\overline{x}_d = 3.625$ and $s_d = 2.066$

$$t = \frac{3.625 - 3}{\dfrac{2.066}{\sqrt{8}}} = .86$$

Since the calculated t of 0.86 does not exceed the t critical value 3.0, the null hypothesis is not rejected. There is not sufficient evidence to support the conclusion that the mean number of words recalled after 1 hour exceeds the mean number of words recalled after 24 hours by more than 3.

12.21   Let μ_d denote the true mean difference of Peptide secretion of
         children with autistic syndrome. (Before - After)

         H_0: $\mu_d = 0$ H_a: $\mu_d > 0$

         The test statistic is: the signed-rank sum.

         With n = 10 and α = 0.01, reject H_0 if the signed-rank sum is
         greater than or equal to 45.

         The differences and signed-ranks are:
         Differences: 5, 7, 13, 15, 16, 22, 31, 55, 58, 77
         Signed-ranks: 1, 2, 3, 4, 5, 6, 7, 8, 9, 10
         Signed-rank sum = 55

         Since the signed-rank sum of 55 exceeds the critical value of
         45, the null hypothesis is rejected. There is sufficient
         evidence to conclude that the restricted diet is successful in
         reducing Peptide secretion.

12.23   Let μ_d denote the true mean difference in epinephrine
         concentration for the two anesthetics. (isoflurance minus
         halothane)

         H_0: $\mu_d = 0$ H_a: $\mu_d \neq 0$

         The test statistic is: the signed-rank sum.

         With n = 10 and α = 0.05, reject H_0 if the signed-rank sum $\leq$ -39
         or signed-rank sum $\geq$ 39.

         The differences and signed-ranks are:
         Differences: -.02, .12, .37, .01, .08, -.52, -.07, .18, -.15,
                       -.09
         Signed-ranks: -2, 6, 9, 1, 4, -10, -3, 8, -7, -5
         Signed-rank sum = 1

         Since the calculated value of the test statistic does not fall
         into the rejection region, H_0 is not rejected. The data suggests
         that there is not any difference in the mean epinephrine
         concentration for the two anesthetics. The assumption that must
         be made about the epinephrine concentration distributions is
         that they are identical with respect to shape and spread, so
         that if they do differ, they differ only with respect to the
         location of their centers.

12.25   Let μ_1 denote the true mean phosphate content when using the
         Gravimetric technique and μ_2 the true mean phosphate content
         when using the Spectrophotometric technique.

         The differences are: -.3, 2.8, 3.9, .6, 1.2, -1.1.

The pairwise averages for these differences are:

difference

	-1.1	-.3	.6	1.2	2.8	3.9
-1.1	-1.1	-.7	-.25	.05	0.85	1.40
-.3		-.3	.15	.45	1.25	1.80
.6			.60	.90	1.70	2.25
1.2				1.20	2.00	2.55
2.8					2.80	3.35
3.9						3.90

difference

Arranging the pairwise averages in order yields

-1.1 -.7 -.2 -.05 .15 .45 .60 .85 .90 1.2
1.25 1.4 1.7 1.80 2.00 2.25 2.55 2.80 3.35 3.9

With n = 6 and confidence level of 95%, Appendix Table IX yields a d value equal to 1. Counting in 1 value from each end of the ordered pairwise averages yields an approximate 95% confidence interval of (-1.1, 3.9) for μ_d. Thus, with 95% confidence, the difference in mean phosphate content as determined by the two procedures is between -1.1 and 3.9 units.

12.27 a) Let μ_d denote the true mean difference in time for entry to first stroke. (hole minus flat)

$H_0: \mu_d = 0$ $H_a: \mu_d \neq 0$

The test statistic is: the signed-rank sum.

With n = 10 and α = 0.01 (actual value α = .02), reject H_0 if the signed-rank sum $\leq$ -45 or if the signed-rank sum $\geq$ 45.

The differences and signed ranks are:
Differences: .12, -.13, .11, -.07, -.17
Signed-ranks: 5, -6, 3.5, -2, -8
Differences: .14, .18, -.25, -.01, -.11
Signed-ranks: 7, 9, -10, -1, -3.5
Signed-rank sum = -6

Since the calculated value of -6 does not fall in the rejection region, H_0 is not rejected. Thus, the data suggests that there is no difference in mean time from entry to first stroke for the two entry methods.

b) Let μ_d denote the true mean difference in initial velocity. (hole minus flat)

$H_0: \mu_d = 0$ $H_a: \mu_d \neq 0$

The test statistic is: the signed-rank sum.

With n = 9 and α = .05, reject H_0 if the signed-rank sum $\leq$ -33 or if the signed-rank sum $\geq$ 33.

(Note: n = 9 rather than 10, since one of the differences is 0 and is therefore removed from the analysis.)

The differences and signed ranks are:
Differences: -1.1, .1, -2.4, -1.0, -3.0, -0.9, -1.4, -1.7, 1.2
Signed-ranks: -4, 1, -8, -3, -9, -2, -6, -7, 5
Signed-rank sum = -33

Since the calculated value of the signed-rank sum of -33 is less than or equal to the critical value -33, H_0 is rejected. Thus, the data does suggest that there is a difference in the true mean initial velocity of the two entry methods.

12.29   Let μ_d denote the true mean difference in lung capacity. (post-operative minus pre-operative)

H_0: μ_d = 0 H_a: μ_d > 0

The test statistic is: $z = \dfrac{\text{signed-rank sum}}{\dfrac{\sqrt{n(n+1)(2n+1)}}{6}}$

For α = 0.05, reject H_0 if z > 1.645.

The differences and signed ranks are:
Differences: 80, 340, 350, 100, 640, -115, 545, 220, 630, 800, 120

Signed-ranks: 5, 13, 14, 6, 20, -8, 16, 11, 19, 21, 9
Differences: 240, -20, 880, 570, 40, -20, 580, 130, 70, 450, 110

Signed-ranks: 12, -1.5, 22, 17, 3, -1.5, 18, 10, 4, 15, 7
Signed-rank sum = 231

$$z = \frac{231}{\dfrac{\sqrt{22(23)(45)}}{6}} = \frac{231}{61.6} = 3.75$$

Since the calculated value of z exceeds the z critical 1.645, H_0 is rejected. The data does suggest that surgery increases the mean lung capacity.

12.31   Let μ denote the true mean processing time.

H_0: μ = 2 H_a: μ > 2

The test statistic is: the signed-rank sum.

With n = 10 and α = 0.05, H_0 if signed-rank sum $\geq$ 33.

Process time	Process time minus 2	Signed ranks
1.4	-.6	-8.5
2.1	.1	2
1.9	-.1	-2
1.7	-.3	-6
2.4	.4	7
2.9	.9	10
1.8	-.2	-4.5
1.9	-.1	-2
2.6	.6	8.5
2.2	.2	4.5

signed-rank sum = 9

Since the calculated value of the signed-rank sum does not fall into the rejection region, the null hypothesis is not rejected. There is insufficient data to conclude that the mean processing time exceeds two minutes.

12.33   Let μ_d denote the mean difference in reaction time. (before minus after)

H_0: $\mu_d = 0$ H_a: $\mu_d < 0$

The test statistic is:

$$t = \frac{\overline{x}_d - 0}{\frac{s_d}{\sqrt{n}}} \quad \text{with d.f.} = 6$$

For $\alpha = .05$, reject H_0 if $t < -1.94$.

The differences are: $-.1$, 0, $-.2$, $-.1$, 0, $.1$, $-.2$.

From these: $\overline{x}_d = -0.0714$ and $s_d = 0.1113$

$$t = \frac{-.0714 - 0}{\frac{.1113}{\sqrt{7}}} = -1.698$$

Since the calculated t of -1.698 does not fall into the rejection region, the null hypothesis is not rejected. The data does not suggest that 2 oz. of alcohol increases mean reaction time.

12.35   a)   Let μ_d denote the true mean difference in heart rate. (trained minus untrained)

H_0: $\mu_d = 0$ H_a: $\mu_d \neq 0$

The test statistic is:

$$t = \frac{\overline{x}_d - 0}{\frac{s_d}{\sqrt{n}}} \quad \text{with d.f.} = 7$$

For $\alpha = 0.05$, reject H_0 if $t > 2.37$ or if $t < -2.37$.

The differences are: -5, 10, -5, 55, 13, 25, -15, -10.

From these: $\overline{x}_d = 8.5$ and $s_d = 23.04$

$$t = \frac{8.5 - 0}{\frac{23.04}{\sqrt{8}}} = 1.04$$

Since the calculated t value does not fall into the rejection region, H_0 is not rejected. There is not sufficient evidence to indicate a difference between trained and untrained children with respect to mean resting heart rate.

b) Let μ_d denote the true mean difference in cardiac output. (trained minus untrained)

H_0: $\mu_d = 0$ H_a: $\mu_d \neq 0$

The test statistic is:

$$t = \frac{\overline{x}_d - 0}{\dfrac{s_d}{\sqrt{n}}} \quad \text{with d.f.} = 7$$

For α = 0.05, reject H_0 if t > 2.37 or if t < -2.37.

The differences are: .3, .5, .8, 4.6, 1.2, -.3, 0, .4.

From these: $\overline{x}_d$ = .9375 and s_d = 1.549

$$t = \frac{.9375 - 0}{\dfrac{1.549}{\sqrt{8}}} = 1.71$$

Since the calculated t of 1.71 does not fall into the rejection region, H_0 is not rejected. There is not sufficient evidence to suggest that mean resting cardiac output differs for trained and untrained children.

c) The researchers used paired samples to control and account for factors which might affect the characteristics under study.

12.37   Let μ_d denote the true mean difference in net earnings between 1982 and 1983. (1983 minus 1982)

H_0: $\mu_d = 0$ H_a: $\mu_d > 0$

The test statistic is:

$$t = \frac{\overline{x}_d - 0}{\dfrac{s_d}{\sqrt{n}}} \quad \text{with d.f.} = 9$$

For α = 0.01, reject H_0 if t > 2.82.

The differences are: 56,0, 8.1, -44.8, 23.2, -3.4, 19.7, -28.7, 17.4, 6.0, -8.0.

From these: $\overline{x}_d$ = 4.55 and s_d = 28.22

$$t = \frac{4.55 - 0}{\dfrac{28.22}{\sqrt{10}}} = 0.51$$

Since the calculated t value of 0.51 does not fall into the rejection region, H_0 is not rejected. There is not sufficient evidence to indicate that mean net earnings increased from 1982 to 1983.

Section 13.1

13.1   a) $y = -5.0 + .017x$

      b)   When $x = 1000$, $y = -5 + .017(1000) = 12$
           When $x = 2000$, $y = -5 + .017(2000) = 29$

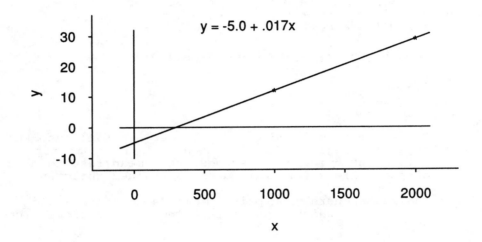

      c)   When $x = 2100$, $y = -5 + (.017)(2100) = 30.7$

      d) $.017$

      e) $.017(100) = 1.7$

      f)   When $x = 500$, $y = -5 + .017(500) = 3.5$
The model does not yield a usage value that is
unreasonable (that is, negative). There is no information
to suggest the model is adequate or inadequate for houses
of this size.

13.3   a)   The mean value of serum manganese when Mn intake is 4.0 is
$-2 + 1.4(4) = 3.6$.

           The mean value of serum manganese when Mn intake is 4.5 is
$-2 + 1.4(4.5) = 4.3$.

      b) $\dfrac{5 - 3.6}{1.2} = 1.17$

           P(serum Mn over 5) = $P(1.17 < z) = 1 - .8790 = .121$

      c)   The mean value of serum manganese when MN intake is 5 is
$-2 + 1.4(5) = 5$.

$$\frac{5 - 5}{1.2} = 0, \quad \frac{3.8 - 5}{1.2} = -1$$

P(serum Mn over 5) = P(0 < z) = .5
P(serum Mn below 3.8) = P(z < -1) = .1587

13.5   a) $y = \alpha + \beta x$ is the equation of the population regression line and $\hat{y} = a + bx$ is the equation of the least squares line (the estimated regression line).

       b)   The quantity b is a statistic. It is the slope of the estimated regression line. The quantity β is a population characteristic. It is the slope of the population regression line. The quantity b is an estimate of β.

       c) $\alpha + \beta x^*$ is the true mean y-value for $x = x^*$. As such $\alpha + \beta x^*$ is a population characteristic. $a + bx^*$ is a point estimate of the mean y value when $x = x^*$ or $a + bx^*$ is a point estimate of an individual y value to be observed when $x = x^*$. The quantity $a + bx^*$ is a statistic.

13.7   a) $r^2 = 1 - \dfrac{SSResid}{SSTo} = 1 - \dfrac{27.890}{73.937} = 1 - .3772 = 0.6228$

       b) $s_e^2 = \dfrac{SSResid}{n - 2} = 1 - \dfrac{27.890}{13 - 2} = \dfrac{27.890}{11} = 2.5355$

          $s_e = \sqrt{2.5355} = 1.5923$

          The magnitude of a typical deviation of observed residence half-time (y) from the least-squares line is about 1.59 hours.

       c)   b = 3.4307

       d) $\hat{y} = .0119 + 3.4307(1) = 3.4426$

13.9   a) $r^2 = 1 - \dfrac{2620.57}{22398.05} = 0.883$

       b) $s_e = \sqrt{\dfrac{2620.57}{14}} = \sqrt{187.184} = 13.682$ with 14 d.f.

13.11   a)

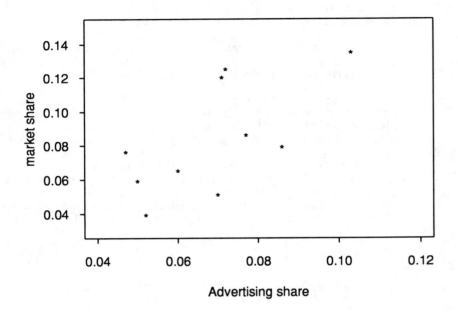

There seems to be a general tendency for y to increase at a constant rate as x increases. However, there is also quite a bit of variability in the y values. It is questionable whether a simple linear regression will adequately describe the relationship between x and y.

b)   Summary values are: $n = 10$, $\Sigma x = 0.688$, $\Sigma x^2 = 0.050072$, $\Sigma y = .0835$, $\Sigma y^2 = .079491$, $\Sigma xy = 0.060861$.

$$b = \frac{0.060861 - \left[\frac{(.688)(.835)}{10}\right]}{0.050072 - \left[\frac{(.688)^2}{10}\right]} = \frac{.003413}{.0027376} = 1.246712$$

$a = .0835 - (1.246712)(.0688) = -.002274$

The equation of the estimated regression line is
$\hat{y} = -.002274 + 1.246712x$.

The predicted market share, when advertising share is .09, would be $-.002274 + 1.246712(.09) = .10993$.

c) $SSTo = .079491 - \left[\frac{(.835)^2}{10}\right] = .0097685$

$SSResid = .079491 - (-.002274)(.835) - (1.246712)(.060861)$
$= .0055135$

$$r^2 = 1 - \frac{.0055135}{.0097685} = 1 - .564 = .436$$

This means that 43.6% of the total variability in market share (y) can be explained by the simple linear regression model relating market share and advertising share (x).

d) $s_e = \sqrt{\dfrac{.0055135}{8}} = \sqrt{.000689} = .0263$ with 8 d.f.

13.13 a)   σ represents the standard deviation of the random
            deviation e. It is the typical deviation of the errors
            about the true regression line. σ_b represents the standard
            deviation of the statistic b. It is the typical deviation
            of b from its mean value β.

       b) s_b is an estimate of σ_b based on the sample data. It is an
            estimate of the typical deviation of b from its mean value
            β.

13.15 For n = 9 observations in Exercise 4.32, $\Sigma(x - \overline{x})^2 =$
       $70.5 - \left[\frac{(24)^2}{9}\right] = 6.5$. For the n = 8 observations given here,
       $\Sigma x = 22$, $\Sigma x^2 = 69$, and $\Sigma(x - \overline{x})^2 = 69 - \left[\frac{(22)^2}{8}\right] = 8.5$. Since
       $\sigma_b = \frac{\sigma}{\sqrt{\Sigma(x - \overline{x})^2}}$, σ_b would be smaller for these eight x values,
       than for the nine x values given in Exercise 4.32. Hence, if the
       objective had been to estimate β as accurately as possible, it
       would be better to use the data set with these 8 x-values.

13.17 a) $r^2 = 1 - \frac{561.46}{2401.85} = 0.766$

       b) $s_e = \sqrt{\frac{561.46}{13}} = \sqrt{43.189} = 6.5719$

            $\Sigma x^2 - \left[\frac{(\Sigma x)^2}{n}\right] = 13.92 - \left[\frac{(14.1)^2}{15}\right] = 0.666$

            $s_b = \frac{6.5719}{\sqrt{0.666}} = \frac{6.5719}{.8161} = 8.0529$

       c) $\Sigma xy - \left[\frac{(\Sigma x)(\Sigma y)}{n}\right] = 1387.2 - \left[\frac{(14.1)(1438.5)}{15}\right] = 35.01$

            $b = \frac{35.01}{0.666} = 52.567$

            The 90% confidence interval for β is
            52.567 ± (1.77)(8.0529) = 52.567 ± 14.254
                   = (38.313, 66.821).

13.19 a)   From the computer output the P-value for testing H_0: β = 0
            against H_a: β ≠ 0 is found to be 0.000. Hence the null
            hypothesis would be rejected for any reasonable α.
            Therefore there does appear to be a useful linear
            relationship between average wage and quit rate.

       b)   A confidence interval for β is required. The confidence
            interval for β based on this sample data is
            -0.34655 ± (2.16)(.05866) = -.034655 ± .12671
                   = (-0.161365, .092055)

13.21 Summary values are: n = 15, $\Sigma x = 609$, $\Sigma x^2 = 28037$, $\Sigma y = 33.1$,
       $\Sigma y^2 = 84.45$, $\Sigma xy = 1156.8$.

$$b = \frac{1156.8 - \left[\frac{(609)(33.1)}{15}\right]}{28037 - \left[\frac{(609)^2}{15}\right]} = \frac{-187.06}{3311.60} = -0.0564863$$

$$a = \frac{[33.1 - (-.0564863)(609)]}{15} = 4.5$$

SSResid = 84.45 - (4.5)(33.1) - (-.0564863)(1156.8) = .843

$$s_e^2 = \frac{.843}{13} = .06485$$

$$s_b^2 = \frac{.06485}{3311.6} = .0000196, \quad s_b = .004425$$

The 99% confidence interval for β is:
-.056 ± (3.01)(.004425) = -.056 ± .013 = (-.069, -.043).

13.23   Summary values are: n = 9, $\Sigma x = 220$, $\Sigma x^2 = 6544$, $\Sigma y = 736$, $\Sigma y^2 = 61524$, $\Sigma xy = 17084$.

a) $$b = \frac{17084 - \left[\frac{(220)(736)}{9}\right]}{6544 - \left[\frac{(220)^2}{9}\right]} = \frac{-907.11}{1166.22} = -.77782$$

$$a = \frac{[(736) - (-.77782)(220)]}{9} = 100.79116$$

b)   Let β denote the expected change in survival percentage associated with a one unit increase in ammonia exposure.

H_0: $\beta = 0$ H_a: $\beta \neq 0$

The test statistic is: $t = \frac{b}{s_b}$ with d.f. = 7.

For significance level .05, reject H_0 if t < -2.37 or if t > 2.37.

SSResid = 61524 - (100.79116)(736) - (-.77782)(17084)
       = 629.98312

$$s_e^2 = \frac{629.98312}{7} = 89.99759$$

$$s_b^2 = \frac{89.99759}{1166.22} = .0772, \quad s_b = .2778$$

$$t = \frac{-.778}{.2778} = -2.80$$

Since the t calculated value of -2.80 falls in the rejection region, the null hypothesis is rejected. The data suggests that the simple linear regression model is useful for predicting survival percentage from knowledge of ammonia exposure.

c)   When ammonia exposure is 30, the predicted survival percentage would be 100.79 - .778(30) = 77.45.

d)   The 90% confidence interval for β is:
$$-.77782 \pm (1.90)(.2778) = -.77782 \pm .528$$
$$= (-1.306, -.250).$$

13.25   Let β denote the average change in milk pH associated with a one unit increase in temperature.

H_0: $\beta = 0$ H_a: $\beta < 0$

The test statistic is: $t = \dfrac{b}{s_b}$ with d.f. = 14.

At level of significance .01, reject H_0 if $t < -2.62$.

Computations: n = 16, Σx = 678, Σy = 104.54,

$$\Sigma xy - \left[\frac{(\Sigma x)(\Sigma y)}{n}\right] = 4376.36 - \left[\frac{(678)(104.54)}{16}\right] = -53.5225,$$

$$\Sigma x^2 - \left[\frac{(\Sigma x)^2}{n}\right] = 36056 - \left[\frac{(678)^2}{16}\right] = 7325.75.$$

$$b = \frac{-53.5225}{7325.75} = -.0073$$

$$a = 6.53375 - (-.0073)(42.375) = 6.8431$$

$$SSResid = 683.447 - 6.8431(104.54) - (-.0073)(4376.36)$$
$$= .016754$$

$$s_e = \sqrt{\frac{.016754}{14}} = \sqrt{.001197} = .0346$$

$$s_b = \frac{.0346}{\sqrt{7325.75}} = .000404$$

$$t = \frac{-.0073}{.000404} = -18.07$$

Since the calculated t of -18.07 is less than the critical t of -2.62, H_0 is rejected. There is sufficient evidence in the sample to conclude that there is a negative (inverse) linear relationship between temperature and pH.

13.27   Let β denote the expected change in cranial capacity associated with a one unit increase in chord length.

H_0: $\beta = 20$ H_a: $\beta \neq 20$

The test statistic is: $t = \dfrac{b - 20}{s_b}$ with d.f. = 5.

At level of significance .05, reject H_0 if $t < -2.57$ or if $t > 2.57$.

$n = 7$, $\Sigma x = 569$, $\Sigma x^2 = 46375$, $\Sigma y = 6310$, $\Sigma y^2 = 5764600$, $\Sigma xy = 515660$

$$b = \frac{515660 - \left[\dfrac{(569)(6310)}{17}\right]}{46375 - \left[\dfrac{(569)^2}{7}\right]} = \frac{2747.14}{123.43} = 22.25694$$

$$a = \frac{[6310 - (22.25694)(569)]}{7} = -907.74269$$

$$SSResid = 5764600 - (-907.74269)(6310) - (22.25694)(515660)$$
$$= 15442.6935$$

$$s_e^2 = \frac{15442.6935}{5} = 3088.5387$$

$$s_b^2 = \frac{3088.5387}{123.43} = 25.0226, \quad s_b = 5.002$$

$$t = \frac{(22.25694 - 20)}{5.002} = 0.45$$

Since the t calculated value of 0.45 does not fall in the rejection region, the null hypothesis is not rejected. This new experimental data does not contradict the prior belief that β is 20.

13.29   A confidence interval is an interval estimate of a population mean, whereas a prediction interval is an interval of plausible values for the single future observation.

A 95% prediction interval is one for which 95% of all possible samples would yield interval limits capturing the future observation.

13.31   a) $s_{a+b(2)} = 16.486 \sqrt{\dfrac{1}{20} + \dfrac{(2 - 2.5)^2}{25}} = 16.486 \sqrt{.06} = 4.038$

b)   Since 3 is the same distance from $\overline{x} = 2.5$ as 2 is, then $s_{a+b(2)} = s_{a+b(3)}$. Hence, $s_{a+b(3)} = 4.038$.

c) $s_{a+b(2.8)} = 16.486 \sqrt{\dfrac{1}{20} + \dfrac{(2.8 - 2.5)^2}{25}} = 3.817$

d) $s_{a+bx}*$ is smallest when $x* = \overline{x}$. Hence, for this data set $s_{a+bx}*$ is smallest when $x* = 2.5$.

13.33   a)   The 95% prediction interval for an observation to be made when $x* = 40$ would be

$6.5511 \pm 2.15 \sqrt{(.0356)^2 + (.008955)^2} =$
$6.5511 \pm 2.15(.0367) = 6.5511 \pm .0789 = (6.4722, 6.6300)$.

b)   The 99% prediction interval for an observation to be made when $x* = 35$ would be

$6.5876 \pm 2.98 \sqrt{(.0356)^2 + (.009414)^2}$
$6.5876 \pm 2.98(.0368) = 6.5876 \pm .1097 = (6.4779, 6.6973)$.

c)   Yes, because $x* = 60$ is farther from the mean value of x, which is 42.375, than is 40 or 35.

13.35   a) $b = \dfrac{57760 - \left[\dfrac{(1350)(600)}{15}\right]}{155400 - \left[\dfrac{(1350)^2}{15}\right]} = \dfrac{3760}{33900} = .1109$

$a = 40 - (.1109)(90) = 30.019$

The equation for the estimated regression line is:
$\hat{y} = 30.019 + .1109x$.

b)   When $x = 100$, the point estimate of $\alpha + \beta(100)$ is:
$30.019 + .1109(100) = 41.109$.

SSResid $= 24869.33 - (30.019)(600) - (.1109)(57760)$
          $= 452.346$

$s_e^2 = \dfrac{452.346}{13} = 34.7958$

$s_{a+b(100)}^2 = 34.7958\left[\dfrac{1}{15} + \dfrac{(100 - 90)^2}{33900}\right] = 2.422$

$S_{a+b(100)} = 1.5564$

The 90% confidence interval for the mean blood level for people who work where the air lead level is 100 is:

$41.109 \pm (1.77)(1.5564) = 41.109 \pm 2.755$
$$= (38.354, 43.864).$$

c) The prediction interval is:

$41.109 \pm (1.77)\sqrt{34.7958 + 2.422} = 41.109 \pm 10.798$
$$= (30.311, 51.907).$$

d) The interval of part (b) is for the mean blood level of all people who work where the air lead level is 100. The interval of part (c) is for a single randomly selected individual who works where the air lead level is 100.

13.37 a)
$$b = \frac{1081.5 - \left[\frac{(269)(51)}{14}\right]}{7445 - \left[\frac{(269)^2}{14}\right]} = \frac{101.571}{2276.357} = .04462$$

$a = 3.6429 - (.04462)(19.214) = 2.78551$

The equation of the estimated regression line is:
$\hat{y} = 2.78551 + .04462x$.

b) $H_0: \beta = 0 \qquad H_a: \beta \neq 0$

The test statistic is: $t = \dfrac{b}{s_b}$ with d.f. = 12.

For significance level .05, reject H_0 if $t < -2.18$ or if $t > 2.18$.

$SSResid = 190.78 - (2.78551)(51) - (.00462)(1081.5)$
$$= .46246$$

$s_e^2 = \dfrac{.46246}{12} = .0385$

$s_b^2 = \dfrac{.0385}{2276.357} = .0000169, s_b = .004113$

$t = \dfrac{.04462}{.004113} = 10.85$

Since the t calculated value of 10.85 falls in the rejection region, the null hypothesis is rejected. The data suggests that the simple linear regression model provides useful information for predicting moisture content from knowledge of time.

c) The point estimate of the moisture content of an individual box that has been on the shelf 30 days is:
$2.78551 + .04462(30) = 4.124$.

The 95% prediction interval is:

$$4.124 \pm (2.18)\sqrt{.0385}\sqrt{1 + \frac{1}{14} + \frac{(30 - 19.214)^2}{2276.357}}$$

$$4.124 \pm 2.18(.2079) = 4.124 \pm .453 = (3.671,\ 4.577).$$

d) Since 4.1 is in the interval constructed in (c), it is very plausible that a box of cereal that has been on the shelf 30 days will not be acceptable.

13.39 Summary values are: $n = 8$, $\Sigma x = 157$, $\Sigma x^2 = 4401$, $\Sigma y = 253$, $\Sigma y^2 = 8299$, $\Sigma xy = 4381$.

a)
$$b = \frac{4381 - \left[\frac{(157)(253)}{8}\right]}{4401 - \left[\frac{(157)^2}{8}\right]} = \frac{-584.125}{1319.875} = -.44256$$

$$a = \frac{[253 - (-.44256)(157)]}{8} = 40.31024$$

The equation of the estimated regression line is:
$\hat{y} = 40.31024 - .44256x$.

b) SSResid $= 8299 - (40.31024)(253) - (-.44256)(4381)$
$= 39.365$

$$s_e^2 = \frac{39.365}{6} = 6.5608$$

$$s_{a+b(18)}^2 = 6.5608\left[\frac{1}{8} + \frac{(18 - 19.625)^2}{1319.875}\right] = .8332$$

$$s_{a+b(18)} = .9128$$

$$a+b(18) = 40.31 - .44256(18) = 32.344$$

The 95% confidence interval for the mean number of rhizobia per seed for seeds stored 18 weeks is:
$32.344 \pm (2.45)(.9128) = 32.344 \pm 2.236$
$= (30.108,\ 34.58)$.

c) When $x^* = 22$, $a + b(22) = 40.31 - .44256(22) = 30.574$.

$$s_{a+b(22)}^2 = 6.5608\left[\frac{1}{8} + \frac{(22 - 19.625)^2}{1319.875}\right] = .8481$$

$$s_{a+b(22)} = .9209$$

The 95% confidence interval for the mean number of rhizobia per seed for seeds stored 22 weeks is:
$30.574 \pm (2.45)(.9209) = 30.574 \pm 2.256 = (28.32,\ 32.83)$.

13.41   a)   When $x^* = 200$, $a + bx^* = 2.142 + (.0068)(200) = 3.502$.

             The 95% prediction interval for a single observation to be made on extraction time when pressure is 200 is

$$3.502 \pm 2.31 \sqrt{(1.06)^2 + (.0384)^2} = 3.502 \pm 2.31(.1127)$$
$$= 3.502 \pm .2604 = (3.2416, 3.7624)$$

  b)   The 99% prediction interval for a single observation to be made on extraction time when pressure is 250 is

$$[2.142 + (.0068)(250)] \pm 3.36 \sqrt{(.106)^2 + (.0349)^2}$$
$$= 3.842 \pm 3.36(.1116) = 3.842 \pm .375 = (3.467, 4.217).$$

13.43   Summary values are: $n = 6$, $\Sigma x = 496$, $\Sigma x^2 = 41272$, $\Sigma y = 69.6$, $\Sigma y^2 = 942.28$, $\Sigma xy = 5942.6$

SSTo = 134.92, SSResid = 2.29

$b = .70173$, $a = -46.41$, $s_e^2 = .5725$

Let $\alpha + \beta(82)$ denote the true average number of hours of chiller operation when maximum outdoor temperature is 82.

H_0: $\alpha + \beta(82) = 12$ H_a: $\alpha + \beta(82) < 12$

The test statistic is: $t = \dfrac{[a + b(82)] - 12}{s_{a+b(82)}}$ with d.f. = 4.

For significance level 0.01, reject H_0 if $t < -3.75$.

$a + b(82) = -46.41 + .70173(82) = 11.132$

$$s_{a+b(82)}^2 = .5725\left[\frac{1}{6} + \frac{(82 - 82.667)^2}{269.333}\right] = .0963$$

$s_{a+b(82)} = .3104$

$$t = \frac{(11.132 - 12)}{.3104} = -2.80$$

Since the t calculated value of -2.80 does not fall into the rejection region, the null hypothesis is not rejected. The data suggests that the true average number of hours of chiller operation is not less than 12 when maximum outdoor temperature is 82. The manufacturer is advised not to produce this system.

13.45 The quantity r is a statistic as its value is calculated from the sample. It is a measure of how strongly the sample x and y values are related. The value of r is an estimate of ρ. The quantity ρ is a population characteristic. It measures the strength of association between the x and y values in the population.

13.47 Let ρ denote the true correlation between KCL extractable aluminum and the amount of lime required to bring soil pH to 7.0.

H_0: $\rho = 0$ $\qquad$ H_a: $\rho \neq 0$

$$t = \frac{r}{\sqrt{\frac{(1 - r^2)}{(n-2)}}} \qquad \text{with d.f. = 22}$$

For significance level .01, reject H_0 if t < -2.82 or if t > 2.82.

$$\Sigma xy - \left[\frac{(\Sigma x)(\Sigma y)}{n} \right] = 658.455 - \left[\frac{(48.15)(263.5)}{24} \right] = 129.808$$

$$\Sigma x^2 - \left[\frac{(\Sigma x)^2}{n} \right] = 155.4685 - \left[\frac{(48.15)^2}{24} \right] = 58.8676$$

$$\Sigma y^2 - \left[\frac{(\Sigma y)^2}{n} \right] = 3750.53 - \left[\frac{(263.5)^2}{24} \right] = 857.5196$$

$$r = \frac{129.808}{\sqrt{(58.8676)(857.5196)}} = .5778$$

$$t = \frac{.5778}{\sqrt{\frac{[1 - (.5778)^2]}{22}}} = 3.32$$

Since the calculated t of 3.32 exceeds the critical t of 2.82, H_0 is rejected. The data supports the conclusion that there is a correlation between KCL extractable aluminum and amount of lime required to bring soil pH to 7.0.

13.49 From the summary quantities:

$$\Sigma xy - \left[\frac{(\Sigma x)(\Sigma y)}{n} \right] = 673.65 - \left[\frac{(136.02)(39.35)}{9} \right] = 78.94$$

$$\Sigma x^2 - \left[\frac{(\Sigma x)^2}{n} \right] = 3602.65 - \left[\frac{(136.02)^2}{9} \right] = 1546.93$$

$$\Sigma y^2 - \left[\frac{(\Sigma y)^2}{n} \right] = 184.27 - \left[\frac{(39.35)^2}{9} \right] = 12.223$$

$$r = \frac{78.94}{\sqrt{(1546.93)(12.223)}} = \frac{78.94}{137.51} = .574$$

Section 13.4

a) Let ρ denote the correlation between surface and subsurface concentration.

$H_0: \rho = 0 \qquad H_a: \rho \neq 0$

$$t = \frac{r}{\sqrt{\frac{(1 - r^2)}{(n-2)}}} \quad \text{with d.f.} = 7$$

For significance level 0.05, reject H_0 if $t < -2.37$ or if $t > 2.37$.

$$t = \frac{.574}{\sqrt{\frac{[1 - (.574)^2]}{7}}} = 1.85$$

Since the t calculated value of 1.85 does not fall into the rejection region, the null hypothesis is not rejected. The data does not support the conclusion that there is a linear relationship between surface and subsurface concentration.

b) From Appendix Table IV, $.20 > \text{P-value} > .10$.

13.51 Let ρ denote the correlation between memory size and retail prices.

$H_0: \rho = 0 \qquad H_a: \rho \neq 0$

$$t = \frac{r}{\sqrt{\frac{(1 - r^2)}{(n-2)}}} \quad \text{with d.f.} = 11$$

For significance level .10, reject H_0 if $t < -1.80$ or if $t > 1.80$.

The summary statistics are: $n = 13$, $\Sigma x = 412$, $\Sigma x^2 = 28974$, $\Sigma y = 5873$, $\Sigma y^2 = 3850617$, $\Sigma xy = 286648$,

$$\Sigma xy - \left[\frac{(\Sigma x)(\Sigma y)}{n}\right] = 100519.08, \quad \Sigma x^2 - \left[\frac{(\Sigma x)^2}{n}\right] = 15916.769,$$

$$\Sigma y^2 - \left[\frac{(\Sigma y)^2}{n}\right] = 1197376.3.$$

$$r = \frac{100519.08}{\sqrt{(15916.769)(1197376.3)}} = .728$$

$$t = \frac{.728}{\sqrt{\frac{[1 - (.728)^2]}{11}}} = 3.52$$

Since the t calculated value of 3.52 falls in the rejection region, the null hypothesis is rejected. The data suggests that there is a linear relationship between memory size and retail price.

13.53   The summary statistics are: n = 13, Σx = 6.55, Σx^2 = 3.5775, Σy = 1397, Σy^2 = 152283, Σxy = 725.35,

$\Sigma xy - \left[\dfrac{(\Sigma x)(\Sigma y)}{n}\right]$ = 21.477, $\Sigma x^2 - \left[\dfrac{(\Sigma x)^2}{n}\right]$ = .2773,

$\Sigma y^2 - \left[\dfrac{(\Sigma y)^2}{n}\right]$ = 2159.23.

a) $r = \dfrac{21.477}{\sqrt{(.2773)(2159.23)}}$ = .8777

b)   Let ρ denote the correlation between muskrat's increase in swimming speed and the increase in the sweep arc of hind feet.

H_0: ρ = 0 H_a: ρ > 0

$t = \dfrac{r}{\sqrt{\dfrac{(1 - r^2)}{(n-2)}}}$ with d.f. = 11

For significance level .05, reject H_0 if t > 1.8.

$t = \dfrac{.8777}{\sqrt{\dfrac{[1 - (.8777)^2]}{11}}}$ = 6.07

Since the t calculated value of 6.07 falls in the rejection region, the null hypothesis is rejected. The data suggests that there is a useful linear relationship between the increase in muskrat swimming speed and the increase in the sweep arc of hind feet.

c)   From Appendix Table IV, P-value < .0005.

d)   Since a change in the unit of measurement for either variable has no effect on the value of r, the conclusion would remain the same if x were expressed in feet per second instead of meters per second.

13.55 H_0: ρ = 0 H_a: $\rho \neq$ 0

$t = \dfrac{r}{\sqrt{\dfrac{(1 - r^2)}{(n-2)}}}$ with d.f. = 9998

For significance level 0.05, reject H_0 if t < -1.96 or if t > 1.96.

$t = \dfrac{.022}{\sqrt{\dfrac{[1 - (.022)^2]}{(9998)}}}$ = 2.2

Since the t calculated value of 2.2 falls in the rejection region, the null hypothesis is rejected. The results are statistically significant. Because of the extremely large sample size, it is easy to detect a value of ρ which differs from zero by a small amount. If ρ is very close to zero, but not zero, the practical significance of a non-zero correlation may be of little consequence.

13.57   a)   The assumptions required in order that the simple linear
             regression model be appropriate are:

     (i) the observations on vigor be independent of one
          another.

    (ii) the distribution of vigor be normally distributed
          with constant variance at each value of stem
          density.

   (iii) the mean value of vigor is a linear function of stem
          density.

        b)

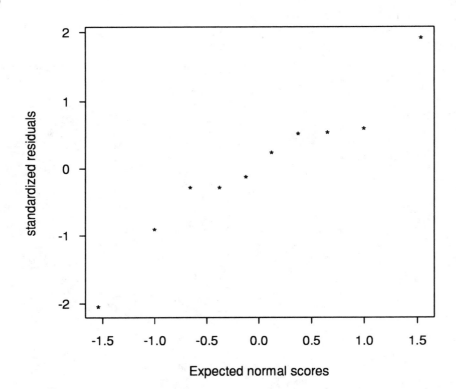

The normal probability plot appears to be reasonably
straight. Hence the assumption that the random deviation
distribution is normal is plausible.

c)

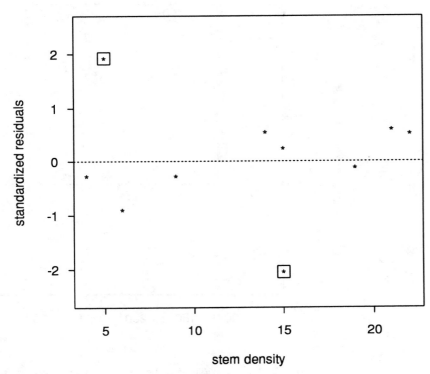

stem density

There are two residuals that are unusually large. The
corresponding points are enclosed in boxes on the graph
above.

d)   Except for the two unusually large residuals, the negative
residuals are associated with small x values, and the
positive residuals are associated with large x values.
This would cause one to question the appropriateness of a
simple linear regression model.

13.59   a)

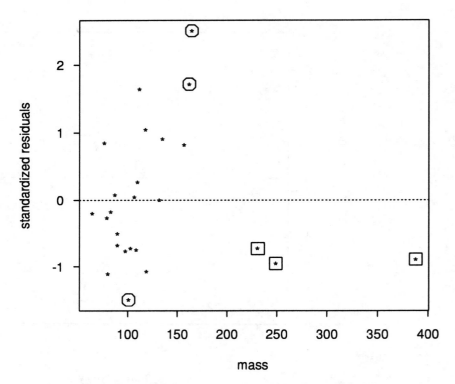

The several large residuals are marked by circles. The potentially influential observations are marked by rectangles.

b)   The residuals associated with the potentially influential observations are all negative. Without these three, there appears to be a positive trend to the plot. The plot suggests that the simple linear regression model might not be appropriate.

c)   There does not appear to be any pattern in the plot that would suggest that it is unreasonable to assume that the variance of y is the same at each x value.

13.61

Year	X	Y	Y-Pred	Residual
1963	188.5	2.26	1.750	0.51000
1964	191.3	2.60	2.478	0.12200
1965	193.8	2.78	3.128	-0.34800
1966	195.9	3.24	3.674	-0.43400
1967	197.9	3.80	4.194	-0.39400
1968	199.9	4.47	4.714	-0.24400
1969	201.9	4.99	5.234	-0.24400
1970	203.2	5.57	5.572	-0.00200
1971	206.3	6.00	6.378	-0.37800
1972	208.2	5.89	6.872	-0.98200
1973	209.9	8.64	7.314	1.32600

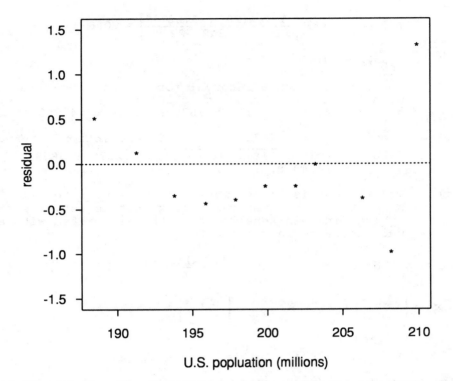

U.S. popluation (millions)

The residuals are positive in 1963 and 1964, then they are
negative from 1965 through 1972, followed by a positive residual
in 1973. The residuals exhibit a pattern in the plot and thus
the plot casts doubt on the appropriateness of the simple linear
regression model.

13.63 a) When x* = 20, a+b(20) = 2.2255 + .1521(20) = 5.2675.

$$S^2_{a+b(20)} = .1187\left[\frac{1}{11} + \frac{(20 - 26.627)^2}{342.622}\right] = .026$$

$$S_{a+b(20)} = .1613$$

The 95% confidence interval for the mean tensile modulus when bound rubber content is 20% is:
5.2675 ± 2.26(.1613) = 5.2675 ± .3645 = (4.903, 5.632)

b) From (a), y = 5.2675 and the 95% prediction interval is
5.2675 ± 2.26 $\sqrt{.1187 + (.1613)^2}$ = 5.2675 ± .8597
= (4.4078, 6.1272)

c) The request is to estimate the "true <u>mean</u> tensile strength" at a given value of x, not to estimate the "<u>change</u> in mean tensile strength" associated with a one-unit increase in x.

13.65 From the data: n = 9, $\Sigma xy - \left[\frac{(\Sigma x)(\Sigma y)}{n}\right]$ = 1389.767,

$\Sigma x^2 - \left[\frac{(\Sigma x)^2}{n}\right]$ = 93.429, $\Sigma y^2 - \left[\frac{(\Sigma y)^2}{n}\right]$ = 295534

$$r = \frac{1389.767}{\sqrt{(93.429)(295534)}} = .2645.$$

Let ρ denote the correlation between eye weight and cornea thickness.

H_0: ρ = 0 H_a: ρ > 0

$$t = \frac{r}{\sqrt{\frac{(1 - r^2)}{(n-2)}}}$$ with d.f. = 7

For significance level 0.05, reject H_0 if t > 1.90.

$$t = \frac{.2645}{\sqrt{\frac{[1 - (.2645)^2]}{7}}} = .73$$

Since the t calculated value of .73 does not fall in the rejection region, the null hypothesis is not rejected. The data does not support the conclusion of a positive correlation between eye weight and cornea thickness.

13.67 a) Let ρ denote the correlation coefficient between soil hardness and trail length.

H_0: ρ = 0 H_a: ρ < 0

$$t = \frac{r}{\sqrt{\frac{(1 - r^2)}{(n-2)}}} \qquad \text{with d.f. = 59}$$

For significance level 0.05, reject H_0 if $t < -1.67$ (approx.).

$$t = \frac{-.6213}{\sqrt{\frac{[1 - (-.6213)^2]}{59}}} = -6.09$$

Since the t calculated value of -6.09 falls in the rejection region, the null hypothesis is rejected. The data supports the conclusion of a negative correlation between trail length and soil hardness.

b) When $x^* = 6$, $a+b(6) = 11.607 - 1.4187(6) = 3.0948$

$$s^2_{a+b(6)} = (2.35)^2\left[\frac{1}{59} + \frac{(6 - 4.5)^2}{250}\right] = .1433$$

$$s_{a+b(6)} = .3786$$

The 95% confidence interval for the mean trail length when soil hardness is 6 is:
$3.0948 \pm 2.00(.3786) = 3.0948 \pm .7572 = (2.3376, 3.8520)$

c) When $x^* = 10$, $a+b(10) = 11.607 - 1.4187(10) = -2.58$

According to the least-squares line, the predicted rail length when soil hardness is 10 is -2.58. Since trail length cannot be negative, the predicted value makes no sense. Therefore, one would not use the simple linear regression model to predict trail length when hardness is 10.

13.69   a)

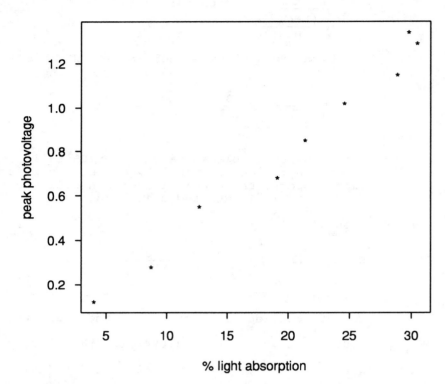

The plot suggests that a straight line model might
adequately describe the relationship between percent light
absorption and peak photovoltage.

b) $\Sigma xy - \left[\dfrac{(\Sigma x)(\Sigma y)}{n}\right] = 178.683 - \left[\dfrac{(179.7)(7.28)}{9}\right] = 33.326$

$\Sigma x^2 - \left[\dfrac{(\Sigma x)^2}{n}\right] = 4334.41 - \left[\dfrac{(179.7)^2}{9}\right] = 746.4$

$\Sigma y^2 - \left[\dfrac{(\Sigma y)^2}{n}\right] = 7.4028 - \left[\dfrac{(7.28)^2}{9}\right] = 1.514$

$b = \dfrac{33.326}{746.4} = .044649$

$a = .809 - .044649(19.667) = -0.08259$

$\hat{y} = -0.08259 + .044649x$

c) $r^2 = \dfrac{(33.326)^2}{(746.4)(1.514)} = .983$

d)   When $x^* = 19.1$, $y = -.08259 + .044649(19.1) = .7702$.
The corresponding residual is $.68 - .7702 = -.0902$.

202

e) $H_0: \beta = 0 \qquad H_a: \beta \neq 0$

The test statistic is: $t = \dfrac{b}{s_b}$ with d.f. = 7

For significance level .05, reject H_0 if $t < -2.37$ or $t > 2.37$.

$b = .044649$

$$s_e^2 = \frac{7.4028 - (-.08259)(7.28) - .044649(178.683)}{7}$$

$$s_e^2 = \frac{.02604}{7} = .00372$$

$$s_b^2 = \frac{.00372}{746.4} = 4.984 \cdot 10^{-6}, \; s_b = .00223$$

$$t = \frac{.044649}{.00223} = 20.00$$

Since the calculated t of 20 exceeds the critical t of 2.37, H_0 is rejected. The data does support the conclusion that there is a useful linear relationship between percent light absorption and peak photovoltage.

f)   In the absence of a specified confidence level, 95% will be used. The 95% confidence interval for the average change in peak photovoltage associated with a 1% increase in light absorption is
$.044649 \pm 2.37(.00223) = .044649 \pm .00529$
$\qquad\qquad = (.039359, .049939)$

g)   In the absence of a specified confidence level, 95% will be used. When $x^* = 20$, $\hat{y} = -.08259 + .044649(20) = .8104$

$$s_{a+b(20)} = .061 \sqrt{\frac{1}{9} + \frac{(20 - 19.667)^2}{746.4}} = .0204$$

The 95% confidence interval of true average peak photovoltage when percent light absorption is 20 is
$.8104 \pm 2.37(.0204) = .8104 \pm .0483 = (.7621, .8587)$.

13.71 a)   The summary values are: $n = 10$, $\Sigma x = 25$, $\Sigma x^2 = 145$, $\Sigma y = -.4$, $\Sigma y^2 = 43.88$, $\Sigma xy = 55.5$.

$$b = \frac{55.5 - \left[\dfrac{(25)(-.4)}{10}\right]}{145 - \left[\dfrac{(25)^2}{10}\right]} = \frac{56.5}{82.5} = .68485$$

$a = -.04 - .68485(2.5) = -1.7521$

The equation of the estimated regression line is:

$\hat{y} = -1.7521 + .68485x$.

b) $SSResid = 43.88 - (-1.7521)(-.4) - (.68485)(55.5)$
$= 5.169985$

$$s_e^2 = \frac{5.169985}{8} = .646248$$

$$s_a^2 = s_{a+b(0)}^2 = s_e^2\left[\frac{1}{n} + \frac{\overline{x}^2}{\sum x^2 - \left[\frac{(\sum x)^2}{n}\right]}\right] = .646248\left[\frac{1}{10} + \frac{(2.5)^2}{82.5}\right]$$

$s_a^2 = .113583$, $s_a = .337$

H_0: $\alpha = 0$ H_a: $\alpha \neq 0$

The test statistic is: $t = \frac{a}{s_a}$ with d.f. = 8.

For significance level .05, reject H_0 it $t < -2.31$ or if $t > 2.31$.

$$t = \frac{-1.7521}{.337} = -5.20$$

Since the t calculated value of -5.20 falls in the rejection region, the null hypothesis is rejected. The data suggests that the y intercept of the true regression line differs from zero.

c) The 95% confidence interval for α is:
$-1.7521 \pm (2.31)(.337) = -1.7521 \pm .7785$
$= (-2.5306, -.9736)$

Since the interval does not contain the value zero, zero is not one of the plausible values for α.

13.73 Summary values for Leptodactylus ocellatus: n = 9, $\sum x = 64.2$, $\sum x^2 = 500.78$, $\sum y = 19.6$, $\sum y^2 = 47.28$, $\sum xy = 153.36$

From these: b = .31636, SSResid = .3099, $\sum(x-\overline{x})^2 = 42.82$

Summary values for Bufa marinus: n = 8, $\sum x = 55.9$, $\sum x^2 = 425.15$, $\sum y = 21.6$, $\sum y^2 = 62.92$, $\sum xy = 163.63$

From these: b = .35978, SSResid = .1279, $\sum(x-\overline{x})^2 = 34.549$

$$s^2 = \frac{.3099 + .1279}{9 + 8 - 4} = \frac{.4378}{13} = .0337$$

H_0: $\beta - \beta'$ H_a: $\beta \neq \beta'$

The test statistic is:

$$t = \frac{b - b'}{\sqrt{\frac{s^2}{SS_x} + \frac{s^2}{SS_x'}}}$$ with d.f. = 9 + 8 - 4 = 13.

For significance level .05, reject H_0 if $t < -2.16$ or if $t > 2.16$.

$$t = \frac{.31636 - .35978}{\sqrt{\dfrac{.0337}{42.82} + \dfrac{.0337}{34.549}}} = \frac{-.04342}{.04198} = -1.03$$

Since the t calculated value of -1.03 does not fall in the rejection region, the null hypothesis of equal regression slopes cannot be rejected. The data suggests that the slopes of the true regression lines for the two different frog populations may be identical.

13.75   When the point is included in the computations, the slope will have a larger negative value than if the point is excluded from the computations. Changing the slope will also have an effect on the intercept.

13.77   Since the P-value of .0076 is smaller than most reasonable levels of significance, the conclusion of the model utility test would be that the percentage raise does appear to be linearly related to productivity.

13.79   a)   The e_i's are the deviations of the observations from the population regression line, whereas the residuals are the deviations of the observations from the estimated regression line.

b)   The simple linear regression model states that $y = \alpha + \beta x + e$. Without the random deviation e, the equation implies a deterministic model, whereas the simple linear regression model is probabilistic.

c)   The quantity b is a statistic. Its value is known once the sample has been collected, and different samples result in different b values. Therefore, it does not make sense to test hypotheses about b. Only hypotheses about a population characteristic can be tested.

d)   If $r = +1$ or -1, then each point falls exactly on the regression line and SSResid would equal zero. A true statement is that SSResid is always greater than or equal to zero.

e)   The sum of the residuals must equal zero. Thus, if they are not all exactly zero, at least one must be positive and at least one must be negative. They cannot all be positive. Since there are some positive and no negative values among the reported residuals, the student must have made an error.

f)   SSTo = $\sum (y - \overline{y})^2$ must be greater than or equal to SSResid = $\sum (y - \hat{y})^2$. Thus, the values given must be incorrect.

CHAPTER 14
MULTIPLE REGRESSION ANALYSIS

Section 14.1

14.1   A deterministic model does not have the random deviation component e, while a probabilistic model does contain such a component.

Let y = total number of goods purchased at a service station which sells only one grade of gas and one type of motor oil

x_1 = gallons of gasoline purchased

x_2 = number of quarts of motor oil purchased.

Then y is related to x_1 and x_2 in a deterministic fashion.

Let y = IQ of a child

x_1 = age of the child

x_2 = total years of education of the parents.

Then y is related to x_1 and x_2 in a probabilistic fashion.

14.3   a) β_1 = .237. This value is the expected change in profit margin associated with a one unit increase in net revenue when the number of branch offices remains fixed.

β_2 = -.0002. This value is the expected change in profit margin associated with a one unit increase in the number of branch offices when the net revenue remains fixed.

b)   .237

c) $1.565 + .237(4) - .0002(6500) = 1.213$

14.5   a) $.69 + .47(2) + .00041(.1) - .72(1.2) + .023(6) = 9.634$

b)   It would increase, since the value of β_2 is positive.

14.7   a)   mean $y = 21.09 + .653x_1 + .0022x_2 - .0206x_1^2 + .00004x_2^2$

b)   May 6 is 16 days after April 20. The values of x_1 and x_2 are 16 and 41180 respectively.

$21.09 + .653(16) + .0022(41180) - .0206(16)^2$
$+ .00004(41180)^2 = 67948.5564$

c)   It is smaller. The value for y when x_1 = 32 and x_2 = 41180 is 67943.1836.

14.9   a)

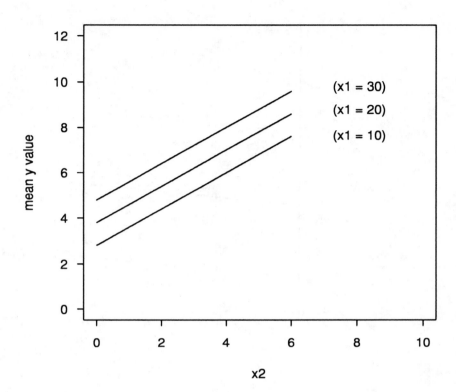

For $x_1 = 30$, $y = 4.8 + .8x_2$
For $x_1 = 20$, $y = 3.8 + .8x_2$
For $x_1 = 10$, $y = 2.8 + .8x_2$

b)

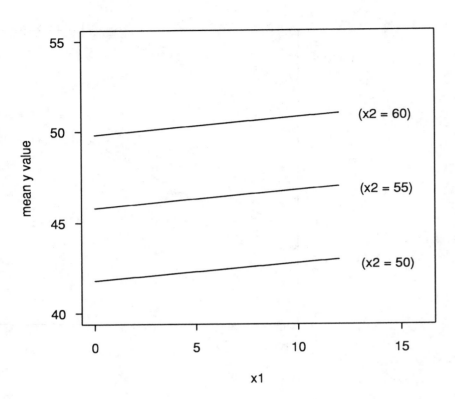

$$\text{For } x_2 = 60, \ y = 49.8 + .1x_2$$
$$\text{For } x_2 = 55, \ y = 45.8 + .1x_2$$
$$\text{For } x_2 = 50, \ y = 41.8 + .1x_2$$

c) The parallel lines in each graph are attributable to the lack of interaction between the two independent variables.

14.11 a) $y = \alpha + \beta_1 x_1 + \beta_2 x_2 + \beta_3 x_3 + e$

b) $y = \alpha + \beta_1 x_1 + \beta_2 x_2 + \beta_3 x_3 + \beta_4 x_1^2 + \beta_5 x_2^2 + \beta_6 x_3^2 + e$

c) $y = \alpha + \beta_1 x_1 + \beta_2 x_2 + \beta_3 x_3 + \beta_4 x_1 x_2 + e$
    $y = \alpha + \beta_1 x_1 + \beta_2 x_2 + \beta_3 x_3 + \beta_4 x_1 x_3 + e$
    $y = \alpha + \beta_1 x_1 + \beta_2 x_2 + \beta_3 x_3 + \beta_4 x_2 x_3 + e$

d) $y = \alpha + \beta_1 x_1 + \beta_2 x_2 + \beta_3 x_3 + \beta_4 x_1^2 + \beta_5 x_2^2 + \beta_6 x_3^2$
    $+ \beta_7 x_1 x_2 + \beta_8 x_1 x_3 + \beta_9 x_2 x_3 + e$

14.13   Three dummy variables would be needed to incorporate a non-
        numerical variable with four categories.

x_3 = 1 if the car is a sub-compact, 0 otherwise

x_4 = 1 if the car is a compact, 0 otherwise

x_5 = 1 if the car is a midsize, 0 otherwise

$y = \alpha + \beta_1 x_1 + \beta_2 x_2 + \beta_3 x_3 + \beta_4 x_4 + \beta_5 x_5 + e$

14.15　a)　The value b_1 = .2 represents the estimated expected change in import volume associated with a one unit increase in expenditure on personal consumption, holding price of import/domestic price constant.

b)　A reduction in domestic price causes the ratio price of import/domestic price to increase. Since the coefficient of the ratio variable x_2 is negative, an increase in x_2 would result in a decrease in the predicted import volume.

c)　The R^2 value of .96 means that 96 percent of the variation in the observed import volume values has been explained by the fitted model.

14.17　H_0: $\beta_1 = \beta_2 = \beta_3 = \beta_4 = \beta_5 = 0$

H_a: At least one of the five β_i's is not zero.

Test statistic: $F = \dfrac{SSRegr / k}{SSResid / [n-(k+1)]}$

With numerator d.f. = 5 and denominator d.f. = 10 - 6 = 4, Appendix Table X gives the level .05 critical value as 6.26. Hence, reject H_0 if F > 6.26.

$F = \dfrac{.170368/5}{.213632/4} = .638$

Since .638 < 6.26, the null hypothesis is not rejected. The data suggests that the independent variables as a group do not provide any information that is useful for predicting soil pH.

14.19　a)　β_3 represents the difference in the mean percent of school board members in the school district who are black for a district in the south compared to a district that is not in the south, when the other independent variables are held constant. b_3 = -4.60 is an estimate of this difference. Since b_3 is negative, the mean percent for districts in the south is 4.60 less than the mean percent for comparable districts that are not in the south.

β_5 represents the change in the mean percent of school board members in a school district who are black associated with a one percent increase in black population in the district, holding the other four independent variables constant. b_5 = 1.07 is an estimate of this change.

b)　H_0: $\beta_4 = \beta_5 = 0$

H_a: At least one of the two β_i's is not zero.

Test statistic: $F = \dfrac{R^2 / k}{(1 - R^2) / [n-(k+1)]}$

With numerator d.f. = 2 and denominator d.f. = 140-(2+1) = 137, Appendix Table X gives the level .01 critical value as approximately 4.77. Hence, reject H_0 if F > 4.77.

$$F = \frac{.730/2}{.270/137} = 185.2$$

Since 185.2 > 4.77, the null hypothesis is rejected. The data suggests that the model has utility.

14.21   a) H_0: $\beta_1 = \beta_2 = \beta_3 = 0$

H_a: At least one of the three β_i's is not zero.

Test statistic: $F = \dfrac{R^2 / k}{(1 - R^2) / [n-(k+1)]}$

With numerator d.f. = 3 and denominator d.f. = 367-4 = 363, Appendix Table X gives the level .05 critical value as 2.60. Hence, reject H_0 if F > 2.60.

$$F = \frac{.16/3}{[1 - (.16)]/363} = 23.05$$

Since 23.05 > 2.60, the null hypothesis is rejected. The data suggests that the fitted model has utility for predicting number of indictments.

b)   The fact that the test results are statistically significant when R^2 = .16 is a bit surprising. However, if n is quite large, even a model for which y is not very strongly related to the predictors will be judged useful by the model utility test. This is analogous to what might happen in the bivariate case if, for example, ρ = .02; with a large enough n, H_0: ρ = 0 will be rejected in favor of the conclusion that there is a positive linear relationship.

14.23   a) H_0: $\beta_1 = \beta_2 = \beta_3 = \beta_4 = 0$

H_a: At least one of the four β_i's is not zero.

Test statistic: $F = \dfrac{SSRegr / k}{SSResid / [n-(k+1)]}$

With numerator d.f. = 4 and denominator d.f. = 38 - 5 = 33, Appendix Table X gives the level .01 critical value as 4.02. Hence, reject H_0 if F > 4.02.

$$F = \frac{1650.02/4}{264.19/33} = 51.53$$

Since 51.53 > 4.02, the null hypothesis is rejected. The data suggests that the fitted regression equation is useful for predicting puffin nest density.

b)   The quantities R^2, s_e and the residuals themselves would help to assess the accuracy of predictions.

c) $$R^2 = \frac{1650.02}{1650.02 + 264.19} = .862$$

$$s_e^2 = \frac{264.19}{33} = 8.006, \quad s_e = 2.83$$

The value of R^2 indicates that 86.2% of the variation in the puffin nest density has been explained by the fitted regression.

The value of s_e means that the typical deviation from the mean value is 2.83.

d) Fitting the equation given in this part of the problem would result in a larger sum of squared residuals. This is because the estimated regression equation is chosen to minimize the sum of squared residuals.

14.25 The F statistic is

$$F = \frac{R^2 / k}{(1 - R^2) / [n-(k+1)]} = \frac{.90/k}{(1 - .9)/[15 - (k+1)]} = \frac{9(14 - k)}{k}$$

Substituting values of k (starting with k = 1), computing the value of F and comparing the calculated value against the critical F value from Appendix Table X, it can be shown that if k is 9 or less the model would be judged to be useful. For k = 9, the calculated F is 5.00 and the critical F is 4.77. For k = 10, the calculated F is 3.60, and the critical F is 5.96. A high R^2 value can often be obtained simply by including a great many predictors in the model, even though the actual population relationship between y and the predictors is weak. To pass the model utility test, a model with a high R^2 based on relatively few predictors is needed. In addition, s_e needs to be small.

14.27 Output from MINITAB is given below:

The regression equation is
* Y = 76.4 - 7.3 X1 + 9.6 X2 - 0.91 X3 + 0.0963 X4
  - 13.5 X1-SQ + 2.80 X2-SQ + 0.0280 X3-SQ -0.000320 X4-SQ
  + 3.75 X1*X2 - 0.750 X1*X3 + 0.142 X1*X4 + 2.00 X2*X3
  - 0.125 C2*C4 + 0.00333 X3*X4

Predictor	Coef	Stdev	t-ratio	p
Constant	76.437	9.082	8.42	0.000
X1	-7.35	10.80	-0.68	0.506
X2	9.61	10.80	0.89	0.387
X3	-0.915	1.068	-0.86	0.404
X4	0.09632	0.09834	0.98	0.342
X1-SQ	-13.452	6.599	-2.04	0.058
X2-SQ	2.798	6.599	0.42	0.677
X3-SQ	0.02798	0.06599	0.42	0.677
X4-SQ	-0.0003201	0.0002933	-1.09	0.291
X1*X2	3.750	8.823	0.43	0.676
X1*X3	-0.7500	0.8823	-0.85	0.408
X1*X4	0.14167	0.05882	2.41	0.028
X2*X3	2.0000	0.8823	2.27	0.038
X2*X4	-0.12500	0.05882	-2.13	0.049
X3*X4	0.003333	0.005882	0.57	0.579

s = 0.3529   R-sq = 88.5%   R-sq(adj) = 78.3%

Section 14.2

Analysis of Variance

SOURCE	DF	SS	MS	F	p
Regression	14	15.2641	1.0903	8.75	0.000
Error	16	1.9926	0.1245		
Total	30	17.2568			

a) The estimated regression equation is marked with an asterisk.

b) H_0: $\beta_1 = \beta_2 = \ldots = \beta_{14} = 0$

   H_a: At least one of the fourteen β_i's is not zero.

   Test statistic: $F = \dfrac{SSRegr / k}{SSResid / [n-(k+1)]}$

   Appendix Table X does not give the level .05 critical value for numerator d.f. = 14 and denominator d.f. = 16. However, one can see by looking at the row in the table for 16 degrees of freedom, that the critical values decrease as the degrees of freedom for the numerator increase. The .05 level critical value is 2.49 when the numerator degrees of freedom is 10, and hence, the .05 level critical value is smaller than 2.49 when the degrees of freedom for the numerator is 14.

   $F = \dfrac{15.2641/14}{1.9926/16} = 8.76$

   Since 8.76 > 2.49, it will also be greater than the critical value for numerator d.f. = 14 and denominator d.f. = 16, and so the null hypothesis is rejected. The data suggests that the fitted model has utility for predicting brightness of finished paper.

c) R^2 = .885. Hence, 88.5% of the variation in the observed brightness readings has been explained by the fitted model.

   SSResid = 1.9926. This is the sum of the squared prediction errors.

   S_e = .3529. This is the typical error of prediction.

14.29　It should be determined that there is a useful estimated regression equation before using it to make predictions. This is the reason it is preferable to perform a model utility test before using an estimated regression equation to make predictions.

14.31　a)　The 95% confidence interval for β_3 is:

$$-9.378 \pm (2.08)(4.356) = -9.378 \pm 9.06$$
$$= (-18.438, -.318)$$

With 95% confidence, the expected change in number of fish associated with a one unit change in sea state, while holding the other variable constant, is estimated to be between -18.438 and -.318.

b)　$$-2.179 \pm (1.72)(1.087) = -2.179 \pm 1.87$$
$$= (-4.049, -.309)$$

14.33　a)　The value .469 is an estimate of the expected change in the mean score of students associated with a one unit increase in the student's expected score holding time spent studying and student's grade point average constant.

b)　$H_0: \beta_1 = \beta_2 = \beta_3 = 0$

H_a: At least one of the three β_i's is not zero.

Test statistic: $F = \dfrac{R^2 / k}{(1 - R^2) / [n-(k+1)]}$

With numerator d.f. = 3 and denominator d.f. = 103, Appendix Table X gives the level .05 critical value as approximately 2.70. Hence, reject H_0 if $F > 2.70$.

$$F = \frac{.686/3}{.314/103} = 75.01$$

Since 75.01 > 2.70, the null hypothesis is rejected. The data suggests that there is a useful linear relationship between exam score and at least one of the three predictor variables.

c)　The 95% confidence interval for β_2 is
$3.369 \pm (1.98)(.456) = 3.369 \pm .903 = (2.466, 4.272)$.

d)　The point prediction would be
$2.178 + .469(75) + 3.369(8) + 3.054(2.8) = 72.856$.

e)　The prediction interval would be
$72.856 \pm (1.98) \sqrt{s_e^2 + (1.2)^2}$.

To determine s_e^2, proceed as follows. From the definition of R^2, it follows that SSResid = $(1-R^2)$SSTo. So SSResid = $(1-.686)(10,200) = 3202.8$. Then

$$s_e^2 = \frac{3202.8}{103} = 31.095.$$

The prediction interval becomes

$$72.856 \pm (1.98) \sqrt{31.095 + (1.2)^2} = 72.856 \pm (1.98)(5.704)$$
$$= 72.856 \pm 11.294 = (61.562, 84.150).$$

14.35 $H_0: \beta_3 = 0$ $H_a: \beta_3 \neq 0$

$t = \dfrac{b_3}{s_{b_3}}$ with d.f. = 363

For significance level .05, reject H_0 if $t < -1.96$ or if $t > 1.96$.

$t = \dfrac{.00002}{.000009} = 2.22$

Since the t calculated of 2.22 exceeds the t-critical value 1.96, the null hypothesis is rejected. The conclusion is that the inclusion of the interaction term is important.

14.37 $H_0: \beta_2 = 0$ $H_a: \beta_2 \neq 0$

The test statistic is: $t = \dfrac{b_2}{s_{b_2}}$ with d.f. = 4.

For significance level 0.05, reject H_0 if $t < -2.78$ or if $t > 2.78$.

$t = \dfrac{-1.7155}{.2036} = -8.42$

Since the t calculated value of -8.42 falls in the rejection region, the null hypothesis is rejected. The quadratic term is important to the model, in addition to the linear term.

14.39   No. See Example 14 and the discussion following that example.

14.41   The point prediction for mean phosphate adsorption when $x_1 = 160$ and $x_2 = 39$ is at the midpoint of the given interval. So the value of the point prediction is $(21.40 + 27.20)/2 = 24.3$. The t-critical value for a 95% confidence interval is 2.23. The standard error for the point prediction is equal to $(27.20 - 21.40)/2(2.23) = 1.30$. The t-critical value for a 99% confidence interval is 3.17. Therefore, the 99% confidence interval would be
$24.3 \pm (3.17)(1.3) = 24.3 \pm 4.121 = (20.179, 28.421).$

14.43   a) $H_0: \beta_1 = \beta_2 = 0$

H_a: At least one of the two β_i's is not zero.

Test statistic: $F = \dfrac{\text{SSRegr} / k}{\text{SSResid} / [n-(k+1)]}$

With numerator d.f. = 2 and denominator d.f. = 7, Appendix Table X gives the level .05 critical value as 4.74. Hence, reject H_0 if $F > 4.74$.

$F = \dfrac{237.52/2}{26.98/7} = 30.81$

Since 30.81 > 4.74, the null hypothesis is rejected. The data suggests that the fitted model is useful for predicting plant height.

b) The .05 level t-critical value is 2.37. From the MINITAB output the t-ratio for b_1 is 6.57, and the t-ratio for b_2 is -7.69. Since both t-ratios (in absolute value) exceed the t-critical value of 2.37, both hypotheses would be rejected. The data suggests that both the linear and quadratic terms are important.

c) The point estimate of the mean y value when x = 2 is:
$\hat{y} = 41.74 + 6.581(2) - 2.36(4) = 45.46$.

The 95% confidence interval is:
$45.46 \pm (2.37)(1.037) = 45.46 \pm 2.46 = (43.0, 47.92)$.

With 95% confidence, the mean height of wheat plants treated with x = 2 ($10^2 = 100$ uM of Mn) is estimated to be between 43 and 47.92 cm.

d) The point estimate of the mean y value when x = 1 is:
$\hat{y} = 41.74 + 6.58(1) - 2.36(1) = 45.96$.

The 90% confidence interval is:
$45.46 \pm (1.9)(1.031) = 45.46 \pm 1.96 = (44.0, 47.92)$.

With 90% confidence, the mean height of wheat plants treated with x = 1 (10 = 10 uM of Mn) is estimated to be between 44 and 47.92 cm.

14.45 One possible way would have been to start with the set of predictor variables consisting of all five variables, along with all quadratic terms, and all interaction terms. Then, use a selection procedure like backward elimination to arrive at the given estimated regression equation.

14.47 The model using the three variables x_3, x_9, x_{10} is the choice of the author. It has an adjusted R^2 which is only slightly smaller than the largest adjusted R^2. This model has two less variables, and hence, two more degrees of freedom for estimation of the variance than does the model with largest adjusted R^2.

14.49 In step 1, the model that includes all four variables was used. In step 2, the model with the variable x_1 (grass cover) deleted was used. In step 3, the model with the two variables x_1 (grass cover) and x_2 (mean soil depth) deleted was used. Neither of the remaining two variables can be removed, so the final model uses the two variables x_3 (angle of slope) and x_4 (distance from cliff edge).

14.51 Adjusted R^2 will be substantially smaller than R^2, when the number of predictor variables, k, is large compared to the number of observations n.

14.53 Multicollinearity might be a problem because most homes differ primarily with respect to the number of bedrooms and bathrooms. Hence, it may be that the value of x_3 (total rooms) is approximately equal to $c + x_1 + x_2$. If so, then multicollinearity will be present.

14.55 a) It would have to be greater than or equal to .723, because if you add variables to a model, R^2 cannot decrease.

b) It would have to be less than or equal to .723, because if you take variables out of a model, R^2 cannot increase.

14.57 a) $H_0: \beta_1 = \beta_2 = \beta_3 = \beta_4 = 0$

         H_a: At least one of the four β_i's is not zero.

         Test statistic: $F = \dfrac{R^2 / k}{(1 - R^2) / [n-(k+1)]}$

         With numerator d.f. = 4 and denominator d.f. = 28, Appendix Table X gives the level .05 critical value as 2.71. Hence, reject H_0 if F > 2.71.

         $F = \dfrac{.69/4}{(1 - .69)/28} = 15.58$

         Since 15.58 > 2.71, the null hypothesis is rejected. The data supports the conclusion that the fitted model is useful in predicting base salary for employees of high tech companies.

    b) $\hat{y} = 2.60 + .125(50) + .893(1) + .057(8) - .014(12)$
            $= 10.031$

    c)   The t-ratios are: 1.953, 6.333, 4.071, -2.8. Since the t-ratio associated with β_1 is less than 2 (in absolute magnitude), the variable x_1 can be deleted.

    d)   The 95% confidence interval is:
         $.057 \pm (2.05)(.014) = .057 \pm .0287 = (.0283, .0857)$.

14.59   First, the model using all four variables was fit. The variable age at loading (x_3) was deleted because it had the t-ratio closest to zero and it was between -2 and 2. Then, the model using the three variables x_1, x_2, and x_4 was fit. The variable time (x_4) was deleted because its t-ratio was closest to zero and was between -2 and 2. Finally, the model using the two variables x_1 and x_2 was fit. Neither of these variables could be eliminated since their t-ratios were greater than 2 in absolute magnitude. The final model then, includes slab thickness (x_1) and load (x_2). The predicted tensile strength for a slab that is 25 cm thick, 150 days old, and is subjected to a load of 200 kg for 50 days is:
$\hat{y} = 13 - .487(25) + .0116(150) = 2.565$.

14.61　a)

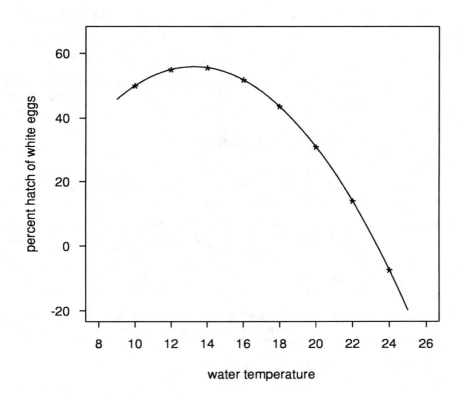

b)　　The claim is very reasonable because 14 is close to where the smooth curve has its highest value.

14.63　a)　　Output from MINITAB is given below.

The regression equation is:
$Y = 1.56 + .0237 X1 - 0.000249 X2.$

Predictor	Coef	Stdev	t-ratio	p
Constant	1.56450	0.07940	19.70	0.000
X1	0.23720	0.05556	4.27	0.000
X2	-0.00024908	0.00003205	-7.77	0.000

s = 0.05330　　　R-sq = 86.5%　　　R-sq(adj) = 85.3%

Analysis of Variance

SOURCE	DF	SS	MS	F	p
Regression	2	0.40151	0.20076	70.66	0.000
Error	22	0.06250	0.00284		
Total	24	0.46402			

b) H_0: β_1 = β_2 = 0

H_a: At least one of the two β_i's is not zero.

Test statistic: $F = \dfrac{\text{SSRegr} / k}{\text{SSResid} / [n-(k+1)]}$

With numerator d.f. = 2 and denominator d.f. = 22, Appendix Table X gives the level .05 critical value as 3.44. Hence, reject H_0 if F > 3.44.

$F = \dfrac{.40151/2}{.0625/22} = 70.67$

Since 70.67 > 3.44, the null hypothesis is rejected. The data suggests that the fitted model is useful for predicting profit margin.

c)   The value for R^2 is .865. This means that 86.5% of the total variation in the observed values for profit margin has been explained by the fitted regression equation. The value for s_e is .0533. This means that the typical deviation from the mean value is .0533, when predicting profit margin using this fitted regression equation.

d)   No. Both variables have associated t-ratios that exceed 2 in absolute magnitude. Hence, neither can be eliminated from the model.

e)

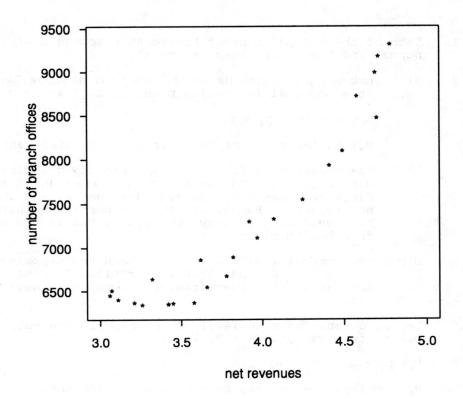

net revenues

There do not appear to be any influential observations.
However, there is substantial evidence of
multicollinearity. The plot shows a pronounced linear
relationship between x_1 and x_2. This is evidence of
multicollinearity between x_1 and x_2.

15.1   Each of the k population of treatment response distributions is
        normal and their variances are equal.

15.3   a)   Let μ_1, μ_2, μ_3 and μ_4 denote the true average length of stay
              in a hospital for health plans 1, 2, 3 and 4 respectively.

              H_0: $\mu_1 = \mu_2 = \mu_3 = \mu_4$

              H_a: At least two of the four μ_i's are different.

        b)   The numerator d.f. = 4 - 1 = 3 and the denominator
              d.f. = 32 - 4 = 28. At α = 0.01, reject H_0 if F > 4.57.
              Since the computed F of 4.37 does not exceed 4.57, H_0 is
              not rejected. Hence, it would be concluded that the
              average length of stay in the hospital is the same for the
              four health plans.

        c)   The numerator d.f. = 4 - 1 = 3 and the denominator
              d.f. = 32 - 4 = 28. Thus the critical F value remains the
              same as in (b). Therefore, the conclusion would be the
              same.

15.5   Let μ_i denote the mean level of chlorophyll concentration for
        plants in variety i (i = 1, 2, 3, 4).

        H_0: $\mu_1 = \mu_2 = \mu_3 = \mu_4$

        H_a: At least two of the four μ_i's are different.

        Test statistic: $F = \dfrac{MSTr}{MSE}$

        Rejection region: Numerator d.f. = k - 1 = 3 and
        denominator d.f. = N - k = 16. For α = .05, Appendix Table X
        gives 3.24 as the approximate critical value. The null
        hypothesis will be rejected if F > 3.24.

        $$\overline{\overline{X}} = \frac{[5(.3) + 5(.24) + 4(.41) + 6(.33)]}{(20)} = .316$$

        $$MSTr = [5(.3 - .316)^2 + 5(.24 - .316)^2 + 4(.41 - .316)^2$$
        $$+ 6(.33 - .316)^2]/3 = .06668/3 = .02223$$

        $$F = \frac{MSTr}{MSE} = \frac{.022223}{.013} = 1.71$$

        Since 1.71 does not exceed the critical F value of 3.24, the
        null hypothesis is not rejected. The data does not suggest that
        true mean chlorophyll concentration differs for the four
        varieties.

15.7   a)   Let μ_i denote the mean satisfaction level for technique i
              (i = 1, 2, 3).

              H_0: $\mu_1 = \mu_2 = \mu_3$
              H_a: At least two of the three μ_i's are different.

              Test statistic: $F = \dfrac{MSTr}{MSE}$

              Rejection region: Numerator d.f. = k - 1 = 2 and
              denominator d.f. = N - k = 30. For α = .05, Appendix Table
              X gives 3.32 as the critical value. The null hypothesis
              will be rejected if F > 3.32.

              $F = \dfrac{MSTr}{MSE} = 4.12$

              Since 4.12 > 3.32, the null hypothesis is rejected.
              The data suggests that true mean satisfaction level
              depends on which interview technique is being used.

        b)   The denominator d.f. = 45 - 3 = 42. The critical value
              decreases as the denominator d.f. increases. Since the
              computed F of 4.12 exceeds the .05 level critical value
              when the denominator d.f. = 30, it will exceed the .05
              level critical value when the denominator d.f. = 42.
              Hence, the null hypothesis would be rejected and the same
              conclusion as in (a) holds.

15.9   Let μ_i denote the mean tensile strength of wire of type i
        (i = 1, 2 ,..., 5).

        H_0: $\mu_1 = \mu_2 = \mu_3 = \mu_4 = \mu_5$
        H_a: At least two of the five μ_i's are different.

        Test statistic: $F = \dfrac{MSTr}{MSE}$

        Rejection region: Numerator d.f. = k - 1 = 4 and denominator
        d.f. = N - k = 15. For α = .01, Appendix Table X gives 4.89 as
        the critical value. The null hypothesis will be rejected if
        F > 4.89.

        $F = \dfrac{MSTr}{MSE} = \dfrac{2573.3}{1394.2} = 1.85$

        Since 1.85 < 4.89, the null hypothesis is not rejected.
        The data suggests that the mean tensile strength is the same for
        the five wire types.

15.11   a)   Let μ_i denote the true mean exam score for teaching method
              i (i = 1, 2, 3, 4, 5).

              H_0: $\mu_1 = \mu_2 = \mu_3 = \mu_4 = \mu_5$
              H_a: At least two of the five μ_i's are different.

              Test statistic: $F = \dfrac{MSTr}{MSE}$

Rejection region: Numerator d.f. = k - 1 = 4 and
denominator d.f. = N - k = 40. For α = .05, Appendix Table
X gives 2.61 as the critical value. The null hypothesis
will be rejected if F > 2.61.

Computations:

$$MSTr = [9(29.3 - 30.82)^2 + 9(28 - 30.82)^2$$
$$+ 9(30.2 - 30.82)^2 + 9(32.4 - 30.82)^2$$
$$+ 9(34.2 - 30.82)^2]/4 = 221.112/4 = 55.278$$

$$MSE = [(4.99)^2 + (5.33)^2 + (3.33)^2 + (2.94)^2$$
$$+ (2.74)^2]/5 = 80.5491/5 = 16.1098$$

$$F = \frac{MSTr}{MSE} = \frac{55.278}{16.1098} = 3.43$$

Since 3.43 > 2.61, the null hypothesis is rejected.
The data suggests that the mean exam score differs for at
least two of the methods of instruction.

b) Let μ_i denote the true mean retention score for teaching
method i (i = 1, 2, 3, 4, 5).

H_0: $\mu_1 = \mu_2 = \mu_3 = \mu_4 = \mu_5$
H_a: At least two of the five μ_i's are different.

Test statistic: $F = \dfrac{MSTr}{MSE}$

Rejection region: Numerator d.f. = k - 1 = 4 and
denominator d.f. = N - k = 40. For α = .05, Appendix Table
X gives 2.61 as the critical value. The null hypothesis
will be rejected if F > 2.61.

Computations:

$$MSTr = [9(30.2 - 29.3)^2 + 9(28.8 - 29.3)^2$$
$$+ 9(26.2 - 29.3)^2 + 9(31.1 - 29.3)^2$$
$$+ 9(30.2 - 29.3)^2]/4 = 132.48/4 = 33.12$$

$$MSE = [(3.82^2 + (5.26)^2 + (4.66)^2 + (4.91)^2$$
$$+ (3.53)^2]/5 = 100.5446/5 = 20.109$$

$$F = \frac{MSTr}{MSE} = \frac{33.12}{20.109} = 1.65$$

Since 1.65 < 2.61, the null hypothesis is not rejected.
The data suggests that the mean retention test score does
not differ for the five teaching methods.

15.13 Let μ_1, μ_2, and μ_3 denote the true mean fog indices for
Scientific America, *Fortune*, and *New Yorker*, respectively.

H_0: $\mu_1 = \mu_2 = \mu_3$
H_a: At least two of the three μ_i's are different.

Test statistic: $F = \dfrac{MSTr}{MSE}$

Rejection region: Numerator d.f. = k - 1 = 2 and denominator d.f. = N - k = 15. For α = .01, Appendix Table X gives 6.36 as the critical value. The null hypothesis will be rejected if F > 6.36.

Computations: $\bar{\bar{x}}$ = 9.666

Magazine	n	Mean	Standard Deviation
S.A.	6	10.968	2.647
F	6	10.68	1.202
N.Y.	6	7.35	1.412

$$MSTr = [6(10.968 - 9.666)^2 + 6(10.68 - 9.666)^2 + 6(7.35 - 9.666)^2]/2 = 48.524/2 = 24.262$$

$$MSE = [(2.647)^2 + (1.202)^2 + (1.412)^2]/3 = 10.443/3 = 3.4812$$

$$F = \frac{MSTr}{MSE} = \frac{24.262}{3.4812} = 6.97$$

Since 6.97 > 6.36, the null hypothesis is rejected. The data suggests that there is a difference between at least two of the mean fog index levels for advertisements appearing in the three magazines.

15.15

Source of Variation	Degrees of Freedom	Sum of Squares	Mean Square	F
Treatments	2	152.18	76.09	5.56
Error	71	970.96	13.675	
Total	73	1123.14		

Let μ_1, μ_2, and μ_3 denote the true mean perceptual incongruities of the depressive group, the functional 'other' group, and the brain-damaged group, respectively.

H_0: $\mu_1 = \mu_2 = \mu_3$

H_a: At least two of the three μ_i's are different.

Test statistic: $F = \dfrac{MSTr}{MSE}$

The .01 level critical value for numerator d.f. = 2 and denominator d.f. = 71 is between 4.98 and 4.79 by Appendix Table X.

Since 5.56 > 4.98, then 5.56 exceeds the .01 level critical value for this analysis. Hence, the null hypothesis is rejected. The data suggests that the mean perception of spacial incongruity is not the same for the three groups.

15.17

Source of Variation	Degrees of Freedom	Sum of Squares	Mean Square	F
Treatments	3	136.14	45.38	5.12
Error	60	532.26	8.871	
Total	63	668.40		

Let μ_i denote the true mean number of breaths per minute for group i (i = 1, 2, 3, 4).

H_0: $\mu_1 = \mu_2 = \mu_3 = \mu_4$
H_a: At least two of the four μ_i's are different.

Test statistic: $F = \dfrac{MSTr}{MSE}$

Rejection region: Numerator d.f. = k - 1 = 3 and denominator d.f. = N - k = 60. For α = .01, Appendix Table X gives 4.13 as the critical value. The null hypothesis will be rejected if F > 4.13.

From the ANOVA table, F = 5.12.

Since 5.12 > 4.13, the null hypothesis is rejected. The data suggests that there is a difference in mean number of breaths per minute for at least two of the four groups.

15.19   Let μ_i denote the true mean DNA content in rats fed a diet containing carbohydrate i (i = 1, 2, 3, 4, 5).

H_0: $\mu_1 = \mu_2 = \mu_3 = \mu_4 = \mu_5$
H_a: At least two of the five μ_i's are different.

Test statistic: $F = \dfrac{MSTr}{MSE}$

Rejection region: Numerator d.f. = k - 1 = 4 and denominator d.f. = N - k = 25. For α = .05, Appendix Table X gives 2.76 as the critical value. The null hypothesis will be rejected if F > 2.76.

Computations:

$SSTr = 6(2.58 - 2.448)^2 + 6(2.63 - 2.448)^2$
$\quad + 6(2.13 - 2.448)^2 + 6(2.41 - 2.448)^2$
$\quad + 6(2.49 - 2.448)^2 = .92928$

Source of Variation	Degrees of Freedom	Sum of Squares	Mean Square	F
Treatments	4	.92928	.23232	2.17
Error	25	2.68072	.10723	
Total	29	3.61		

Since 2.17 < 2.76, the null hypothesis is not rejected. The data suggests that the true average DNA content is not affected by the type of carbohydrate in the diet.

15.21   Let μ_i denote the mean antigen concentration for group i (i = 1, 2, 3).

H_0: $\mu_1 = \mu_2 = \mu_3$
H_a: At least two of the three μ_i's are different.

Test statistic: $F = \dfrac{MSTr}{MSE}$

Rejection region: Numerator d.f. = k - 1 = 2 and denominator d.f. = N - k = 21. For α = .05, Appendix Table X gives 3.47 as the critical value. The null hypothesis will be rejected if F > 3.47.

Source of Variation	Degrees of Freedom	Sum of Squares	Mean Square	F
Treatments	2	1.6297	.8148	4.589
Error	21	3.7289	.1776	
Total	23	5.3586		

Since 4.589 > 3.47, the null hypothesis is rejected. The data suggests that there is a difference in the mean antigen concentrations of at least two of the three groups.

15.23 From Appendix Table XI, the Bonferroni t-value is approximately 2.46. The Bonferroni intervals are:

$\mu_1 - \mu_2$: $(6.6 - 5.37) \pm (2.46) \sqrt{2.028\left(\frac{1}{24} + \frac{1}{24}\right)}$

$= 1.23 \pm 2.46(.411) = 1.23 \pm 1.01 = (.22, 2.24)$

$\mu_1 - \mu_3$: $(6.6 - 5.2) \pm (2.46) \sqrt{2.028\left(\frac{1}{24} + \frac{1}{20}\right)}$

$= 1.40 \pm 2.46(.431) = 1.40 \pm 1.06 = (.34, 2.46)$

$\mu_2 - \mu_3$: $(5.37 - 5.2) \pm (2.46) \sqrt{2.028\left(\frac{1}{24} + \frac{1}{20}\right)}$

$= .17 \pm 2.46(.431) = .17 \pm 1.06 = (-.89, 1.23)$

So μ_1 differs from μ_2 and μ_3, but μ_2 and μ_3 do not differ.

15.25

Group	Simultaneous	Sequential	Control
Mean	$\overline{x}_3$	$\overline{x}_2$	$\overline{x}_1$

15.27 The mean water loss when exposed to 4 hours fumigation is different from all other means. The mean water loss when exposed to 2 hours fumigation is different from that for levels 16 and 0, but not 8. The mean water losses for duration 16, 0, and 8 hours are not different from one another. No other differences are significant.

15.29 Summary quantities:

Brand number	Brand	n	$\overline{x}$	s
1	Imperial	4	14.1	.356
2	Parkay	5	12.8	.430
3	Blue Bonnet	4	13.825	.443
4	Chiffon	4	13.1	.594
5	Mazola	5	17.14	.598
6	Fleishmann's	4	18.1	.648

MINITAB was used to generate the following ANOVA table.

Source of Variation	Degrees of Freedom	Sum of Squares	Mean Square	F
Treatments	5	108.185	21.637	79.264
Error	20	5.460	.273	
Total	25	113.645		

a) Let μ_i denote the mean PAPFUA level for brand i $(i = 1, 2, 3, 4, 5, 6)$.

H_0: $\mu_1 = \mu_2 = \mu_3 = \mu_4 = \mu_5 = \mu_6$
H_a: At least two of the six μ_i's are different.

Test statistic: $F = \dfrac{MSTr}{MSE}$

Rejection region: Since k - 1 = 5 and denominator
d.f. = N - k = 20. For α = .05, Appendix Table X give 2.71
as the critical value. The null hypothesis will be
rejected if F > 2.71.

From the ANOVA table, F = 79.264. Since 79.264 > 2.71, the
null hypothesis is rejected. The data suggests that at
least two of the brands differ with respect to mean PAPFUA
level.

b)   k(k-1)/2 = 6(5)/2 = 15. The Bonferroni t critical value is
3.33.

To compare sample means based on sizes n = 4 and n = 5,
use

$3.33\sqrt{\dfrac{.273}{4} + \dfrac{.273}{5}} = 1.167$.

To compare sample means based on sizes n = 4 and n = 4,
use

$3.33\sqrt{\dfrac{.273}{4} + \dfrac{.273}{4}} = 1.23$.

To compare sample means based on sizes n = 5 and n = 5,
use

$3.33\sqrt{\dfrac{.273}{5} + \dfrac{.273}{5}} = 1.10$.

$\mu_1 - \mu_2$: (14.1 - 12.8) ± 1.167 = (.133, 2.467)

$\mu_1 - \mu_3$: (14.1 - 13.825) ± 1.23 = (-.955, 1.505)

$\mu_1 - \mu_4$: (14.1 - 13.1) ± 1.23 = (-.23, 2.23)

$\mu_1 - \mu_5$: (14.1 - 17.14) ± 1.167 = (-4.207, -1.873)

$\mu_1 - \mu_6$: (14.1 - 18.1) ± 1.23 = (-5.23, -2.77)

$\mu_2 - \mu_3$: (12.8 - 13.825) ± 1.167 = (-2.192, 0.142)

$\mu_2 - \mu_4$: (12.8 - 13.1) ± 1.167 = (-1.467, 0.867)

$\mu_2 - \mu_5$: (12.8 - 17.14) ± 1.10 = (-5.44, -3.24)

$\mu_2 - \mu_6$: (12.8 - 18.1) ± 1.167 = (-6.467, -4.133)

$\mu_3 - \mu_4$: (13.825 - 13.1) ± 1.23 = (-0.505, 1.955)

$\mu_3 - \mu_5$: (13.825 - 17.14) ± 1.167 = (-4.482, -2.148)

$\mu_3 - \mu_6$: (13.825 - 18.1) ± 1.23 = (-5.505, -3.045)

$\mu_4 - \mu_5$: (13.1 - 17.14) ± 1.167 = (-5.207, -2.873)

$\mu_4 - \mu_6$: $(13.1 - 18.1) \pm 1.23 = (-6.23, -3.77)$

$\mu_5 - \mu_6$: $(17.14 - 18.1) \pm 1.167 = (-2.127, .207)$

c)

Brand	2	4	3	1	5	6
Sample Size	5	4	4	4	5	4
Mean	12.8	13.1	13.825	14.1	17.14	18.1

The mean PAPFUA levels for Mazola and Fleishmann's are the same and these two brands are different from the other four. The mean Parkay PAPFUA level differs from that of Imperial but not Chiffon or Blue Bonnet. No other differences are significant.

15.31 A randomized block experiment was used to control the factor
<u>value of house</u>, which definitely affects the assessors'
appraisals. If a completely randomized experiment had been done,
then there would have been danger of having the assessors
appraising houses which were not of similar values. Therefore,
differences between assessors would be partly due to the fact
that the homes were dissimilar, as well as to differences in the
appraisals made.

15.33 a)

Source of Variation	Degrees of Freedom	Sum of Squares	Mean Square	F
Treatments	2	1835.2	917.60	35.62
Blocks	4	93.1	23.28	
Error	8	206.1	25.76	
Total	14	2134.4		

b) H_0: The mean nitrogen concentration does not depend on the
rate of application.

H_a: The mean nitrogen concentration does depend on the
rate of application.

Test statistic: $F = \dfrac{MSTr}{MSE}$

Rejection region: Since $k - 1 = 2$ and $(k - 1)(l - 1) = 8$,
Appendix Table X gives the $\alpha = .05$ level critical value as
4.46. The null hypothesis will be rejected if $F > 4.46$.

From the ANOVA table, $F = 35.62$. Since $35.62 > 4.46$, the
null hypothesis is rejected. The mean nitrogen
concentration does depend on the rate of application.

15.35

Source of Variation	Degrees of Freedom	Sum of Squares	Mean Square	F
Treatments	2	28.78	14.39	1.04
Blocks	17	2977.67	175.16	
Error	34	469.55	13.81	
Total	53	3476.00		

H_0: The mean adaptation score does not depend on the type of
treatment.

H_a: The mean adaptation score does depend on the type of
treatment.

Test statistic: $F = \dfrac{MSTr}{MSE}$

Rejection region: Since $k - 1 = 2$ and $(k - 1)(l - 1) = 34$,
Appendix Table X gives the $\alpha = .05$ level critical value as
between 3.32 and 3.23. The null hypothesis will be rejected if
$F > $ critical F.

From the ANOVA table, F = 1.04. Since 1.04 < 3.23, the null hypothesis is not rejected. The data suggests that the mean adaptation score does not depend on the type of treatment.

15.37   a)   Summary values are: $\overline{x}_1$ = 91.25, $\overline{x}_2$ = 82.75, $\overline{b}_1$ = 92.5, $\overline{b}_2$ = 80, $\overline{b}_3$ = 77.5, $\overline{b}_4$ = 92.5, $\overline{b}_5$ = 88.5, $\overline{b}_6$ = 92.5, $\overline{b}_7$ = 92.5, $\overline{b}_8$ = 80.

SSTr = $8[(91.25)^2 + (82.75)^2] - (8)(2)(87)^2$ = 289

SSB1 = $2[(92.5)^2 + (80)^2 + (77.5)^2 + (92.5)^2 + (88.5)^2$
       $+ (92.5)^2 + (92.5)^2 + (80)^2] - 8(2)(87)^2$ = 623

Source of Variation	Degrees of Freedom	Sum of Squares	Mean Square	F
Treatments	1	289	289.00	1.09
Blocks	7	623	89.00	
Error	7	1858	265.43	
Total	15	2770		

H_0: The mean heart rates of trained and untrained swimmers are the same.

H_a: The mean heart rates of trained and untrained swimmers are not the same.

Test statistic: $F = \dfrac{MSTr}{MSE}$

Rejection region: Since k - 1 = 1 and (k - 1)(1 - 1) = 7, Appendix Table X gives the α = .05 level critical value as 5.59. The null hypothesis will be rejected if F > 5.59.

From the ANOVA table, F = 1.09. Since 1.09 < 5.59, the null hypothesis is not rejected. The data suggests that there is no difference between the mean heart rates of trained and untrained swimmers.

b)   Let μ_1 and μ_2 denote the mean heart rates of trained and untrained swimmers, respectively.

H_0: $\mu_1 - \mu_2 = 0$ H_a: $\mu_1 - \mu_2 \neq 0$

Test statistic: $t = \dfrac{\overline{d}}{s_d/\sqrt{n}}$ with d.f. = 7

For α = .05, reject H_0 if t < -2.37 of if t > 2.37.

The differences are: -5, 10, -5, 55, 13, 25, -15, -10

$\overline{d}$ = 8.5 s_d = 23.04 $t = \dfrac{8.5}{23.04/\sqrt{8}}$ = 1.04

Since the t calculated value of 1.04 does not fall in the rejection region, the null hypothesis is not rejected. The data suggests that there is no difference between the mean heart rates of trained and untrained swimmers.

15.39　H_0: The mean clover accumulation does not depend on the sowing rate.

H_a: The mean clover accumulation does depend on the sowing rate.

Test statistic: $F = \dfrac{MSTr}{MSE}$

Rejection region: Since $k - 1 = 3$ and $(k - 1)(l - 1) = 9$, Appendix Table X gives the $\alpha = .05$ level critical value as 3.86. The null hypothesis will be rejected if $F > 3.86$.

Source of Variation	Degrees of Freedom	Sum of Squares	Mean Square	F
Treatments	3	3141153.5	1040751.17	2.276
Blocks	3	19470550.0	6490183.33	
Error	9	4141165.5	460129.50	
Total	15	26752869.0		

Since the calculated F of 2.276 does not exceed the critical F of 3.86, the null hypothesis is not rejected. The data suggests that the mean clover accumulation does not differ for the different sowing rates.

15.41　a)　H_0: The mean survival rate does not depend on the seed source.

H_a: The mean survival rate does depend on the seed source.

Test statistic: $F = \dfrac{MSTr}{MSE}$

Rejection region: Since $k - 1 = 4$ and $(k - 1)(l - 1) = 12$, Appendix Table X gives the $\alpha = .05$ level critical value as 3.26. The null hypothesis will be rejected if $F > 3.26$.

Source of Variation	Degrees of Freedom	Sum of Squares	Mean Square	F
Treatments	4	4195.603	1048.901	10.932
Blocks	3	336.354	112.118	
Error	12	1151.389	95.949	
Total	19	5683.348		

Since the calculated F of 10.932 exceeds the critical F of 3.26, the null hypothesis is rejected. The data suggests that the mean survival rate does depend on the seed source.

b)   (Bonferroni t critical value) $\sqrt{\dfrac{MSE}{4} + \dfrac{MSE}{4}}$

$= (3.43) \sqrt{\dfrac{95.949}{4} + \dfrac{95.949}{4}} = (3.43)(6.926) = 23.76$

Since each sample is of size 4, the sample means must differ by at least 23.76 before the corresponding population means are judged to be different.

Source	5	2	1	4	3
Mean	51.975	53.525	67.575	77.35	90.25

Source 3 differs in mean survival rate from sources 2 and 5. Source 4 differs from sources 2 and 5. No other differences are significant.

15.43   a)   The plot does suggest an interaction between peer group and self-esteem. The change in average response, when changing from low to high peer group, is not the same for the low self-esteem group and the high self-esteem group. This is indicated by the non-parallel lines.

        b)   The change in the average response is greater for the low self-esteem group than it is for the high self-esteem group, when changing from low to high peer group interaction. Therefore, the authors are correct in their statement.

15.45

Source of Variation	Degrees of Freedom	Sum of Squares	Mean Square	F
Size (A)	1	.088	.088	8.00
Species (B)	2	.048	.024	2.18
Size by Species	2	.048	.024	2.18
Error	12	.132	.011	
Total	17	.316		

H_0: There is no interaction between Size (A) and Species (B).
H_a: There is interaction between Size and Species.

The test statistic is: $F_{AB} = \dfrac{MSAB}{MSE}$.

Rejection region: Numerator d.f. = 2 and denominator d.f. = 12. Appendix Table X gives the .01 level critical value as 6.93. The null hypothesis will be rejected if $F_{AB} > 6.93$.

Since $F_{AB} = 2.18 < 6.93$, the null hypothesis is not rejected. The data suggests that there is no interaction between Size and Species. Hence, hypothesis tests on main effects will be done.

H_0: There are no size main effects.
H_a: There are size main effects.

The test statistic is: $F_A = \dfrac{MSA}{MSE}$.

Rejection region: Numerator d.f. = 1, denominator d.f. = 12. Appendix Table X gives the .01 level critical value as 9.33. The null hypothesis will be rejected if $F_A > 9.33$.

Since $F_A = 8.00 < 9.33$, the null hypothesis of no size main effects is not rejected. The data supports the conclusion that there is no difference between the mean preference indices for the two sizes of bass.

H_0: There are no species main effects.
H_a: There are species main effects.

The test statistic is: $F_B = \dfrac{MSB}{MSE}$.

Rejection region: Numerator d.f. = 2, denominator d.f. = 12. Appendix Table X gives the .01 level critical value as 6.93. The null hypothesis will be rejected if $F_B > 6.93$.

Since $F_B = 2.18 < 6.93$, the null hypothesis of no species main effects is not rejected. The data supports the conclusion that there are no differences between the mean preference indices for the three species of bass.

15.47   The test for no interaction would have numerator d.f. equal to 2 and denominator d.f. equal to 120. Appendix Table X gives the .05 critical value as 3.07. Since $F_{AB} < 1 < 3.07$, the null hypothesis of no interaction is not rejected. Since there appears to be no interaction, hypothesis tests on main effects are appropriate.

The test for no A main effects would have numerator d.f. equal to 2. Appendix Table X gives the .05 critical value as 3.07. Since $F_A = 4.99 > 3.07$, the null hypothesis of no A main effects is rejected. The data suggests that the expectation of opportunity to cheat affects the mean test score.

The test for no B main effects would have numerator d.f. equal to 1. Appendix Table X gives the .05 critical value as 3.92. Since $F_B = 4.81 > 3.92$, the null hypothesis of no B main effects is rejected. The data suggests that perceived payoff affects the mean test score.

15.49

Source of Variation	Degrees of Freedom	Sum of Squares	Mean Square	F
Race	1	857	857	5.57
Sex	1	291	291	1.89
Race by Sex	1	32	32	.21
Error	36	5541	153.92	
Total	39	6721		

a) H_0: There is no interaction between race and sex.
     H_a: There is interaction between race and sex.

Test statistic: $F_{AB} = \dfrac{MSAB}{MSE}$

Rejection region: Numerator d.f. = 1 and denominator d.f. = 36. Appendix Table X gives the .01 level critical value as (approximately) 7.40. The null hypothesis will be rejected if $F_{AB} > 7.40$.

Since $F_{AB} = .21 < 7.40$, the null hypothesis of no interaction between race and sex is not rejected. Thus, hypothesis tests for main effects are appropriate.

b) H_0: There are no race main effects.
      H_a: There are race main effects.

Test statistic: $F_A = \dfrac{MSA}{MSE}$

Rejection region: Numerator d.f. = 1 and denominator
d.f. = 36. Appendix Table X gives the .01 level critical
value as (approximately) 7.40. The null hypothesis will be
rejected if $F_A > 7.40$.

Since $F_A = 5.57 < 7.40$, the null hypothesis of no race
main effects is not rejected. The data suggests that the
true average lengths of sacra do not differ for the two
races.

c) H_0: There are no sex main effects.
      H_a: There are sex main effects.

Test statistic: $F_B = \dfrac{MSB}{MSE}$

Rejection region: Numerator d.f. = 1 and denominator d.f.
= 36. Appendix Table X gives the .01 level critical value
as (approximately) 7.40. The null hypothesis will be
rejected if $F_B > 7.40$.

Since $F_B = 1.89 < 7.40$, the null hypothesis of no sex main
effects is not rejected. The data suggests that the true
average lengths of sacra do not differ for males and
females.

15.51

Source of Variation	Degrees of Freedom	Sum of Squares	Mean Square	F
Rate	2	469.70	234.85	76.14
Soil Type	2	333.94	166.97	54.14
Error	4	12.34	3.08	
Total	8	815.98		

The numerator d.f. = 2 and the denominator d.f. = 4 for each F
test. Appendix Table X gives the .01 level critical value as
18.00. Both computed F ratios exceed the critical value, so it
appears that the total phosphorus uptake depends upon
application rate as well as soil types.

15.53   Let μ_1, μ_2, and μ_3 denote the true mean protoporphyrin levels for the normal workers, alcoholics with siderblasts, and alcoholics without siderblasts, respectively.

H_0: $\mu_1 = \mu_2 = \mu_3$
H_a: At least two of the three μ_i's are different.

Test statistic: Kruskal-Wallis

Rejection region: The number of d.f. for the chi-squared approximation is k - 1 = 2. For α = .05, Appendix Table XII gives 5.99 as the critical value. The null hypothesis will be rejected if KW > 5.99.

Computations:

	Ranks		
	Normal Workers	Alcoholics w/siderblasts	Alcoholics w/o siderblasts
	5	29.5	14.5
	6	35	7.5
	20.5	37	16.5
	10.5	31	19
	16.5	38	20.5
	29.5	39	9
	7.5	36	12
	25	41	4
	27	32	26
	24	34	3
	10.5	40	14.5
	18		1
	23		28
	22		33
	13		2
Sum of ranks	258	392.5	210.5
n_i	15	11	15
$\bar{r}_i$	17.2	35.68	14.03

$$KW = \frac{12}{(41)(42)} \left[15(17.2 - 21)^2 + 11(35.68 - 21)^2 + 15(14.03 - 21)^2 \right] = 23.11$$

Since 23.11 > 5.99, the null hypothesis is rejected. The data strongly suggests that the true mean protoporphrin levels differ for at least two of the three groups.

15.55 H_0: The mean skin potential does not depend on emotions.
H_a: The mean skin potential differs for at least two of the emotions.

Test statistic: Friedman

238

Rejection region: With α = .05 and k - 1 = 3, Appendix Table XII gives the chi-square value as 7.82. The null hypothesis will be rejected if F_r > 7.82.

Computations:

Ranks

Subject (blocks)

Emotion	1	2	3	4	5	6	7	8	$\overline{r}_i$
Fear	4	4	3	4	1	4	4	3	3.375
Happiness	3	2	2	1	4	3	1	4	2.5
Depression	1	3	4	2	3	2	2	2	2.375
Calmness	2	1	1	3	2	1	3	1	1.75

$$F_r = \frac{(12)(8)}{(4)(5)} \; [(3.375 - 2.5)^2 + (2.5 - 2.5)^2 + (2.375 - 2.5)^2 + (1.75 - 2.5)^2] = (4.8)(1.34375) = 6.45$$

Since 6.45 < 7.82, the null hypothesis is not rejected. The data indicates that the true mean skin potential does not depend on type of emotions.

15.57 H_0: The true mean food consumption does not differ for the three experimental conditions.
H_a: The true mean food consumptions differs for at least two of the three experimental conditions.

Test statistic: Friedman

Rejection region: Rejection region: With α = .01 and k - 1 = 2, Appendix Table XII gives the chi-square value as 9.21. The null hypothesis will be rejected if F_r > 9.21.

Computations:

Ranks

Rats (blocks)

Hours	1	2	3	4	5	6	7	8	$\overline{r}_i$
0	1	1	1	1	1	1	1	1	1
24	2	2	2.5	3	2	3	2	2.5	2.375
72	3	3	2.5	2	3	2	3	2.5	2.625

$$F_r = \frac{(12)(8)}{(3)(4)} \; [(1 - 2)^2 + (2.375 - 2)^2 + (2.625 - 2)^2] = 8(1.53125) = 12.25$$

Since 12.25 > 9.21, the null hypothesis is rejected. The data suggests that the true mean food consumption depends on number of hours of food deprivation.

15.59 a) Let μ_i denote the mean energy usage for cooking method i (i = 1, 2, 3, 4, 5).

1) Microwave
2) Pressure Cooker
3) Electric Pressure Cooker
4) Boiling
5) Baking

H_0: $\mu_1 = \mu_2 = \mu_3 = \mu_4 = \mu_5$
H_a: At least two of the five μ_i's are different.

Test statistic: $F = \dfrac{MSTr}{MSE}$

Rejection region: Numerator d.f. = k - 1 = 4 and denominator d.f. = N - k = 10. For α = .05, Appendix Table X gives 3.48 as the critical value. The null hypothesis will be rejected if F > 3.48.

Computations: $\bar{\bar{x}}$ = 562.867, Σx^2 = 6765235, $\bar{x}_1$ = 206.333, $\bar{x}_2$ = 383.333, $\bar{x}_3$ = 394.333, $\bar{x}_4$ = 573.333, $\bar{x}_5$ = 1257

SSTo = 6765235 - 15(562.867)^2 = 2012951.73

SSTr = 3[(206.333)^2 + (383.333)^2 + (394.333)^2 + (573.333)^2 + (1257)^2] - 15(562.867)^2 = 2009047.07

Source of Variation	Degrees of Freedom	Sum of Squares	Mean Square	F
Treatments	4	2009047.07	502261.77	1286.312
Error	10	3904.66	390.47	
Total	14	2012951.73		

From the ANOVA table, F = 1286.312. Since 1286.312 exceeds 3.48, the null hypothesis is rejected. The data strongly suggests that the mean energy usage differs for at least two of the different cooking methods.

b) (Bonferroni critical t value) $\sqrt{\dfrac{MSE}{n_1} + \dfrac{MSE}{n_2}}$

= (3.58) $\sqrt{\dfrac{390.47}{3} + \dfrac{390.47}{3}}$ = (3.58)(16.13) = 57.76

Since each sample is of size 3, the sample means must differ by at least 57.76 before the corresponding population means are judged to be different.

Method	1	2	3	4	5
Mean	206.33	383.33	394.33	573.33	1257

Only methods 2 and 3 are judged to be the same. All other comparisons are judged to be different.

15.61 Let μ_i denote the mean bacteria count for concentration i
(i = 1, 2, 3, 4).

$H_0: \mu_1 = \mu_2 = \mu_3 = \mu_4$
H_a: At least two of the four μ_i's are different.

Test statistic: $F = \dfrac{MSTr}{MSE}$

Rejection region: Numerator d.f. = k - 1 = 3 and denominator
d.f. = N - k = 16. For α = .05, Appendix Table X gives 3.24 as
the critical value. The null hypothesis will be rejected if
F > 3.24.

Computations: $\overline{\overline{x}}$ = 35.9, Σx^2 = 31432, $\overline{x}_1$ = 45.8, $\overline{x}_2$ = 31,
$\overline{x}_3$ = 32.6, $\overline{x}_4$ = 34.2

SSTo = 31432 - 20(35.9)^2 = 5655.8

SSTr = 5[(45.8)^2 + (31)^2 + (32.6)^2 + (34.2)^2] -20(35.9)^2
     = 679

Source of Variation	Degrees of Freedom	Sum of Squares	Mean Square	F
Treatments	3	679.0	226.33	0.73
Error	16	4976.8	311.05	
Total	19	5655.8		

Since 0.73 < 3.24, the null hypothesis is not rejected. The data
suggests that the mean bacteria count is the same for all four
concentration levels.

15.63 The following ANOVA table was produced by using the computer
packaged MINITAB. Because MINITAB does not compute F ratios, the
user must do so from the mean squares which are given. The F
ratios are found by the user dividing the appropriate mean
squares.

SOURCE	DF	SS	MS	F
Oxygen	3	0.1125	0.0375	2.76
Sugar	1	0.1806	0.1806	13.28
Interaction	3	0.0181	0.0060	0.44
Error	8	0.1087	0.0136	
Total	15	0.4200		

Test for interaction:

The .05 level critical value for numerator d.f. = 3 and
denominator d.f. = 8 is 4.07. The F ratio to test for
interaction is F_{AB} = 0.44. Since this value is less than the
critical value, the null hypothesis of no interaction is not
rejected. Thus, it is appropriate to test for main effects.

Supplementary Exercises

Test for oxygen main effects:

The numerator d.f. = 3 and denominator d.f. = 8. Appendix Table X gives the .05 level critical value as 4.07. Since $F_A = 2.76 < 4.07$, the null hypothesis of no oxygen main effects is not rejected. The data suggests that the true average ethanol level does not depend on which oxygen concentration is used.

Test for sugar main effects:

The numerator d.f. = 1 and denominator d.f. = 8. Appendix Table X gives the .05 level critical value as 5.32. Since $F_B = 13.28 > 5.32$, the null hypothesis of no sugar main effects is rejected. The data suggests that the true average ethanol level does differ for the two types of sugar.

15.65   Let μ_i denote the mean strength of mortar type i (i = 1, 2, 3, 4).

$H_0: \mu_1 = \mu_2 = \mu_3 = \mu_4$
H_a: At least two of the four μ_i's are different.

Test statistic: $F = \dfrac{MSTr}{MSE}$

Rejection region: Numerator d.f. = k - 1 = 3 and denominator d.f. = N - k = 8. For α = .05, Appendix Table X gives 4.07 as the critical value. The null hypothesis will be rejected if F > 4.07.

Computations: $\overline{\overline{x}} = 77.255$, $\Sigma x^2 = 96617.14$, $\overline{x}_1 = 33.96$, $\overline{x}_2 = 129.30$, $\overline{x}_3 = 115.84$, $\overline{x}_4 = 29.92$

SSTo = 96617.14 - 12(77.255)^2 = 24997.12

SSTr = $3[(33.96)^2 + (129.30)^2 + (115.84)^2 + (29.92)^2]$ - 12(77.255)^2 = 24937.631

Source of Variation	Degrees of Freedom	Sum of Squares	Mean Square	F
Treatments	3	24937.63	8312.54	1117.28
Error	8	59.49	7.44	
Total	11	24997.12		

Since 1117.28 > 4.07, the null hypothesis is rejected. The data strongly suggests that the mean strength is not the same for the four types of mortar.

(Bonferroni t critical value) $\sqrt{\dfrac{MSE}{n_1} + \dfrac{MSE}{n_2}}$

= (3.48) $\sqrt{\dfrac{7.44}{3} + \dfrac{7.44}{3}}$ = (3.48)(2.227) = 7.75

Since each sample is of size 3, the sample means must differ by at least 7.75 before the corresponding population means are judged to be different.

Mortar	PCM	OCM	RM	PIM
Mean	29.92	33.96	115.84	129.30

The mortar types PCM and OCM are judged to have equal means. The other two types, PIM and RM, are not equal, nor are they the same as PCM and OCM.

15.67   a)

Source of Variation	Degrees of Freedom	Sum of Squares	Mean Square	F
Treatments	4	62222	15555.5	0.97
Error	295	4710021	15966.17	
Total	299	4772243		

Let μ_i denote the mean score of group i on the RSS scale (i = 1, 2, 3, 4, 5).

H_0: $\mu_1 = \mu_2 = \mu_3 = \mu_4 = \mu_5$
H_a: At least two of the five μ_i's are different.

Test statistic: $F = \dfrac{MSTr}{MSE}$

Rejection region: Numerator d.f. = k − 1 = 4 and denominator d.f. = N − k = 295. For α = .05, Appendix Table X gives 2.37 as the critical value. The null hypothesis will be rejected if F > 2.37.

From the ANOVA table, F = 0.97. Since 0.97 < 2.37, the null hypothesis is not rejected. The data does not allow for the conclusion that the mean scores are different for the five groups.

b)   (Bonferroni t critical value) $\sqrt{\dfrac{MSE}{n_1} + \dfrac{MSE}{n_2}}$

$= (2.81) \sqrt{\dfrac{15966.17}{60} + \dfrac{15966.17}{16}} = (2.81)(23.07) = 64.83$

The sample means must differ by at least 64.83 for the corresponding population means to be judged different.

	Time caps worn (%)				
Group	100	75	50	25	0
Mean	82.17	85.19	98.09	102.26	117.05

No differences are judged to be significant. The application of the Bonferroni methods fails to yield conclusions which agree with the statement in the paper.

15.69

Source of Variation	Degrees of Freedom	Sum of Squares	Mean Square	F
Locations	14	.6	.04286	1.89
Months	11	2.3	.20909	9.20
Error	154	3.5	.02273	
Total	179	6.4		

The .05 level critical value is about 1.70 when the numerator d.f. = 14 and denominator d.f. = 154. Since 1.89 > 1.70, the null hypothesis of no location main effects is rejected. The data suggests that the true concentration differs by location.

The .05 level critical value is about 1.79 when the numerator d.f. = 11 and denominator d.f. = 154. Since 9.20 > 1.79, the null hypothesis of no month main effects is rejected. The data suggests that the true mean concentration differs by month of year.

15.71 H_0: The mean concentration of cadmium does not differ for the five levels of depth.

H_a: The mean concentration of cadmium differs for at least two of the five levels of depth.

$$F = \frac{MSTr}{MSE}$$

Rejection region: Since $k - 1 = 4$ and $(k - 1)(l - 1) = 12$, Appendix Table X gives the $\alpha = .01$ level critical value as 5.41. The null hypothesis will be rejected if $F > 5.41$.

Computations: $\overline{\overline{x}} = 1.55$, $\Sigma x^2 = 60.5$, $\overline{x}_1 = 1.75$, $\overline{x}_2 = 1.875$, $\overline{x}_3 = 1.375$, $\overline{x}_4 = 1$, $\overline{x}_5 = 1.75$, $\overline{b}_1 = 1.1$, $\overline{b}_2 = 1.3$, $\overline{b}_3 = 2.2$, $\overline{b}_4 = 1.6$

$SSTo = 60.5 - 20(1.55)^2 = 12.45$

$SSTr = 4[(1.75)^2 + (1.875)^2 + (1.375)^2 + 1^2 + (1.75)^2]$
$\qquad - 20(1.55)^2 = 2.075$

$SSB1 = 5[(1.1)^2 + (1.3)^2 + (2.2)^2 + (1.6)^2] - 20(1.55)^2$
$\qquad = 3.45$

Source of Variation	Degrees of Freedom	Sum of Squares	Mean Square	F
Treatments	4	2.075	0.519	.90
Blocks	3	3.450	1.150	
Error	12	6.925	0.577	
Total	19	12.450		

Since .90 < 5.41, the null hypothesis is not rejected. The data suggests that the mean concentration of cadmium does not differ for the five levels of depth.

15.73 Multiplying each observation in a single-factor ANOVA will change $\overline{x}_i$, $\overline{\overline{x}}$, and s_i by a factor of c. Hence, MSTr and MSE will be also changed, but by a factor of c^2. However, the F ratio remains unchanged because $c^2 MSTr / c^2 MSE = MSTr/MSE$. That is, the c^2 in the numerator and denominator cancel. It is reasonable to expect a test statistic not to depend on upon the unit of measurement.

15.75 Let μ_1, μ_2, and μ_3 denote the mean lifetime for brands 1, 2 and 3 respectively.

H_0: $\mu_1 = \mu_2 = \mu_3$
H_a: At least two of the three μ_i's are different.

Test statistic: $F = \dfrac{MSTr}{MSE}$

Rejection region: Numerator d.f. = 3 - 1 = 2 and denominator d.f. = 21 - 3 = 18. For α = .05, reject H_0 if F > 3.55.

Computations: $\sum x^2 = 45,171$, $\overline{\overline{x}} = 46.1429$, $\overline{x}_1 = 44.571$,
$\overline{x}_2 = 45.857$, $\overline{x}_3 = 48$

SSTo = $45,171 - 21(46.1429)^2 = 45171 - 44712.43 = 458.57$

SSTr = $7(44.571)^2 + 7(45.857)^2 + 7(48)^2 - 44712.43$
$= 44754.43 - 44712.43 = 42$

SSE = SSTo - SSTr = 458.57 - 42 = 416.57

Source of Variation	Degrees of Freedom	Sum of Squares	Mean Square	F
Treatments	2	42	21	.907
Error	18	416.57	23.14	
Total	20	458.57		

Conclusion: The computed F of .907 does not exceed the critical value 3.55, so H_0 is not rejected at level of significance .05. The data does not suggest that there are differences in true mean lifetimes of the three brands of batteries.

15.77

The transformed data is:						$\overline{x}$
Brand 1	3.162	3.742	2.236	3.464	2.828	3.087
Brand 2	4.123	3.742	2.828	3.000	3.464	3.431
Brand 3	3.606	4.243	3.873	4.243	3.162	3.825
Brand 4	3.742	4.690	3.464	4.000	4.123	4.004

SSTo = $264.001 - 20(3.58675)^2 = 6.703$

SSTr = $[5(3.087)^2 + 5(3.431)^2 + 5(3.825)^2 + 5(4.004)^2]$
$- 257.2955 = 259.8199 - 257.2955 = 2.524$

SSE = 6.703 - 2.524 = 4.179

Let μ_1, μ_2, μ_3, μ_4 denote the mean of the square root of the number of flaws for brand 1, 2, 3 and 4 of tape, respectively.

H_0: $\mu_1 = \mu_2 = \mu_3 = \mu_4$
H_a: At least two of the four μ_i's are different.

$$F = \frac{MSTr}{MSE}$$

Numerator d.f. = 3 and denominator d.f. = 16.

At level .01, reject H_0 if $F > 5.29$.

$$F = \frac{2.524/3}{4.179/.16} = 3.22$$

Since the computed F of 3.22 does not exceed the critical F of 5.29, H_0 is not rejected. The data does not suggest that there are differences in true mean square root of the number of flaws for the four brands of tape.

16.1   a)   Reject H_0 if $X^2 > 9.49$.

        b)   Reject H_0 if $X^2 > 13.28$.

        c)   Reject H_0 if $X^2 > 21.67$.

16.3   a)   The critical value would be 16.27. Since the computed value of 19 exceeds the critical value of 16.27, H_0 would be rejected. The data suggests that the percentages of the four types of nuts differ from what the percentages are suppose to be.

        b)   If n = 40, then the chi-square test should not be used since one of the expected cell frequencies would be less than 5.

16.5   a)   Let π_1, π_2, π_3, π_4 denote the true proportions of homicides occurring during Winter, Spring, Summer, and Fall, respectively.

           H_0: $\pi_1 = \pi_2 = \pi_3 = \pi_4 = .25$
           H_a: H_0 is not true.

           Test statistic:

$$X^2 = \sum \frac{(\text{observed count - expected count})^2}{\text{expected count}}$$

           Rejection region: For $\alpha = 0.05$ and d.f. = 3, reject H_0 if $X^2 > 7.82$.

           Computations: n = 1361

Season	Winter	Spring	Summer	Fall
Frequency	328	334	372	327
Expected	340.25	340.25	340.25	340.25

$$X^2 = \frac{(328 - 340.25)^2}{340.25} + \frac{(334 - 340.25)^2}{340.25} + \frac{(372 - 340.25)^2}{340.25}$$
$$+ \frac{(327 - 340.25)^2}{340.25} = .4410 + .1148 + 2.9627 + .5160$$
$$= 4.03$$

           Since 4.03 < 7.82, the null hypothesis is not rejected. The data collected suggests that there is no difference in the proportion of homicides occurring in the four seasons.

        b)   Since the chi-square critical value for 3 degrees of freedom with level .10 is 6.25, and the calculated chi-square value is 4.03, which is smaller than the critical value, one can conclude that the P-value is in excess of .10.

Section 16.1

16.7 a) Let π_i denote the proportion for phenotype i (i=1, 2, 3).

H_0: $\pi_1 = .25$, $\pi_2 = .5$, $\pi_3 = .25$
H_a: H_0 is not true.

Test statistic:

$$X^2 = \sum \frac{(\text{observed count - expected count})^2}{\text{expected count}}$$

Rejection region: For d.f. = 2, Appendix Table XII gives the .05 level critical value as 5.99. The null hypothesis will be rejected if $X^2 > 5.99$.

Computations: The computed X^2 value is 4.63, which is less than 5.99. The null hypothesis is not rejected. The data does not contradict the researcher's theory.

b) For 2 degrees of freedom, the .05 chi-square value is 5.99 and the .10 chi-square value is 4.61. Since the calculated value is between 4.61 and 5.99, one can conclude .10 > P-value > .05.

c) The analysis and conclusion would remain the same. The sample size is used only to calculate the expected cell frequencies. It has no influence on the degrees of freedom, or the critical value.

16.9 Let π_i denote the proportion of homing pigeons who prefer direction i (i = 1, 2, 3, 4, 5, 6, 7, 8).

H_0: $\pi_1 = \frac{1}{8}$, $\pi_2 = \frac{1}{8}$, $\pi_3 = \frac{1}{8}$, $\pi_4 = \frac{1}{8}$, $\pi_5 = \frac{1}{8}$, $\pi_6 = \frac{1}{8}$, $\pi_7 = \frac{1}{8}$, $\pi_8 = \frac{1}{8}$

H_a: H_0 is not true.

Test statistic: $X^2 = \sum \frac{(\text{observed count - expected count})^2}{\text{expected count}}$

Rejection region: For d.f. = 7, Appendix Table XII gives the .10 level critical value as 12.02. The null hypothesis will be rejected if $X^2 > 12.02$.

Computations:

Direction	1	2	3	4	5	6	7	8	Total
Frequency	12	16	17	15	13	20	17	10	120
Expected	15	15	15	15	15	15	15	15	

$$X^2 = \frac{(12 - 15)^2}{15} + \frac{(16 - 15)^2}{15} + \frac{(17 - 15)^2}{15} + \frac{(15 - 15)^2}{15}$$

$$+ \frac{(13 - 15)^2}{15} + \frac{(20 - 15)^2}{15} + \frac{(17 - 15)^2}{15} + \frac{(10 - 15)^2}{15} = \frac{72}{15} = 4.8$$

Since 4.8 < 12.02, the null hypothesis is not rejected. The data supports the hypothesis that when homing pigeons are disoriented in a certain manner, they exhibit no preference for any direction of flight after take-off.

16.11 a)   d.f. = (4 - 1)(5 - 1) = 12. Reject H_0 if $X^2 > 21.03$.

       b)   At α = .10, H_0 would be rejected if $X^2 > 18.55$. With $X^2 = 7.2$, then the null hypothesis would not be rejected. The data would suggest that educational level and preferred candidate are independent factors.

       c)   d.f. = (4 - 1)(4 - 1) = 9. Reject H_0 if $X^2 > 16.92$. With $X^2 = 14.5$, then the null hypothesis would not be rejected. The data would suggest that educational level and preferred candidate are independent factors.

16.13 H_0: Job satisfaction and teaching level are independent.

       H_a: Job satisfaction and teaching level are dependent.

       Test statistic: $X^2 = \sum \dfrac{\text{(observed count - expected count)}^2}{\text{expected count}}$

       Rejection region: For d.f. = 2, Appendix Table XII gives the .05 level critical value as 5.99. The null hypothesis will be rejected if $X^2 > 5.99$.

       Computations:

		Job satisfaction		
		Satisfied	Unsatisfied	
Teaching level	College	74 (63.763)	43 (53.237)	117
	High School	224 (215.270)	171 (179.730)	395
	Elementary	126 (144.967)	140 (121.033)	266
	Total	424	354	778

$$X^2 = \frac{(74 - 62.763)^2}{63.763} + \frac{(43 - 53.237)^2}{53.237} + \frac{(224 - 215.270)^2}{215.270}$$
$$+ \frac{(171 - 179.730)^2}{179.730} + \frac{(126 - 144.967)^2}{144.967} + \frac{(140 - 121.023)^2}{121.023}$$

$$= 1.644 + 1.968 + .354 + .424 + 2.482 + 2.972 = 9.844$$

Since the calculated chi-square value of 9.844 exceeds the chi-square critical value of 5.99, H_0 is rejected. The data supports the conclusion that there is an association between job satisfaction and teaching level.

16.15 H_0: Proportions falling into each group is the same for males and females.

       H_a: H_0 is not true.

       Test statistic: $X^2 = \sum \dfrac{\text{(observed count - expected count)}^2}{\text{expected count}}$

Rejection region: For d.f. = 4, Appendix Table XII gives the .01 level critical value as 13.28. The null hypothesis will be rejected if $X^2 > 13.28$.

Computations:

Experience	1 - 3	4 - 6	7 - 9	10 - 12	13	Total
Male	202 (285.56)	369 (409.83)	482 (475.94)	361 (347.04)	811 (706.63)	2225
Sex						
Female	230 (146.44)	251 (210.17)	238 (244.06)	164 (177.96)	258 (362.37)	1141
Total	432	620	720	525	1069	3366

$$X^2 = \frac{(202 - 285.56)^2}{285.56} + \frac{(369 - 409.83)^2}{409.83} + \frac{(482 - 475.94)^2}{475.94}$$
$$+ \frac{(361 - 347.04)^2}{347.04} + \frac{(811 - 706.63)^2}{706.63} + \frac{(230 - 146.44)^2}{146.44}$$
$$+ \frac{(251 - 210.17)^2}{210.17} + \frac{(238 - 244.06)^2}{244.06} + \frac{(164 - 177.96)^2}{177.96}$$
$$+ \frac{(258 - 362.37)^2}{362.37}$$

$$= 24.452 + 4.068 + 0.077 + 0.562 + 15.415 + 47.682$$
$$+ 7.934 + 0.151 + 1.096 + 30.059 = 131.496$$

Since the calculated chi-square value of 131.496 exceeds the chi-square critical value of 13.28, H_0 is rejected. The proportions falling into each age group is not the same for males and females.

16.17   a) H_0: Student marijuana use is independent of parental drug and alcohol use.

           H_a: Student marijuana use and parental drug and alcohol use are dependent.

Test statistic:

$$X^2 = \sum \frac{(\text{observed count} - \text{expected count})^2}{\text{expected count}}$$

Rejection region: For d.f. = 4 Appendix Table XII gives the .01 level critical value as 13.28. The null hypothesis will be rejected if $X^2 > 13.28$.

Parental use of alcohol and drugs	Student level of marijuana use			Total
	Never	Occasional	Regular	
Neither	141 (119.3)	54 (57.6)	40 (58.1)	235
One	68 (82.8)	44 (39.9)	51 (40.3)	163
Both	17 (23.9)	11 (11.5)	19 (11.6)	47
Total	226	109	110	445

$$X^2 = \frac{(141 - 119.3)^2}{119.3} + \frac{(54 - 57.6)^2}{57.6} + \frac{(40 - 58.1)^2}{58.1}$$
$$+ \frac{(68 - 82.8)^2}{82.8} + \frac{(44 - 39.9)^2}{(39.9)} + \frac{(51 - 40.3)^2}{40.3}$$
$$+ \frac{(17 - 23.9)^2}{23.9} + \frac{(11 - 11.5)^2}{11.5} + \frac{(19 - 11.6)^2}{11.6}$$

$$= 3.947 + .225 + 5.639 + 2.645 + .421 + 2.841 + 1.992$$
$$+ .022 + 4.721 = 22.45$$

Since the calculated chi-square value of 22.45 exceeds the chi-square critical value of 13.28, H_0 is rejected. There does appear to be an association between student use of marijuana and parental use of alcohol and drugs.

b) The chi-square value for 4 d.f. and α = .001 is 18.47. Since the calculated value of 22.45 exceeds 18.47, it follows that the P-value is less than .001.

16.19 H_0: The true proportion of responses from regular viewers falling in the five response categories are the same for the two programs.

H_a: The true proportion of responses from regular viewers falling in the five response categories are not the same for the two programs.

Test statistic: $X^2 = \sum \frac{(\text{observed count} - \text{expected count})^2}{\text{expected count}}$

Rejection region: For d.f. = (2 - 1)(5 - 1) = 4. At significance level .05, reject $X^2 > 9.49$.

Section 16.2

Computations:

| | Series | | |
Response	Three's Company	Hello Larry	Totals
Extremely interesting	19 (11.2)	9 (16.8)	28
Very interesting	24 (18.4)	22 (27.6)	46
Fairly interesting	30 (30.4)	46 (45.6)	76
Not very interesting	10 (18.8)	37 (28.2)	47
Not at all interesting	17 (21.2)	36 (31.8)	53
Totals	100	150	250

$$X^2 = \frac{(19-11.2)^2}{11.2} + \frac{(24-18.4)^2}{18.4} + \frac{(30-30.4)^2}{30.4}$$
$$+ \frac{(10-18.8)^2}{18.8} + \frac{(17-21.2)^2}{21.2} + \frac{(9-16.8)^2}{16.8}$$
$$+ \frac{(22-27.6)^2}{27.6} + \frac{(46-45.6)^2}{45.6} + \frac{(37-28.2)^2}{28.2}$$
$$+ \frac{(36-31.8)^2}{31.8}$$

$$= 5.432 + 1.704 + .005 + 4.119 + .832 + 3.621 + 1.136$$
$$+ .004 + 2.746 + .555 = 20.155$$

Since 20.155 > 9.49, H_0 is rejected. The data suggests that the true proportion of responses from regular viewers falling in the five response categories are not the same for the two programs.

16.21 H_0: The true proportion of people who have had some college education is the same for each of the countries examined.

H_a: The true proportion of people who have had some college education is not the same for each of the countries examined.

Test statistic: $X^2 = \sum \frac{(\text{observed count} - \text{expected count})^2}{\text{expected count}}$

Rejection region: For d.f. = $(5-1)(2-1) = 4$. At significance level .05, reject H_0 if $X^2 > 9.49$.

Section 16.2

Computations:

	Some College Education	No College Education	Total
U.S.	320 (193)	680 (807)	1000
East Ger.	173 (193)	827 (807)	1000
Canada	172 (193)	828 (807)	1000
Sweden	155 (193)	845 (807)	1000
Japan	145 (193)	855 (807)	1000
Total	965	4035	5000

$$X^2 = \frac{(320-193)^2}{193} + \frac{(173-193)^2}{193} + \frac{(172-193)^2}{193}$$
$$+ \frac{(155-193)^2}{193} + \frac{(145-193)^2}{193} + \frac{(680-807)^2}{807}$$
$$+ \frac{(827-807)^2}{807} + \frac{(828-807)^2}{807} + \frac{(845-807)^2}{807}$$
$$+ \frac{(855-807)^2}{807}$$

$X^2 = 83.570 + 2.073 + 2.285 + 7.482 + 11.938 + 19.986$
$+ .496 + .546 + 1.789 + 2.855 = 133.02$

Since 133.02 > 9.49, the null hypothesis is rejected. The data suggests that the true proportion of people who have had some college education is not the same for the five countries studied.

16.23　H_0: The true proportion of those rearrested is the same for each of the three states.

H_a: The true proportion of those rearrested is not the same for each of the three states.

Test statistic: $X^2 = \sum \frac{(\text{observed count - expected count})^2}{\text{expected count}}$

Rejection region: For d.f. = (3 - 1)(2 - 1) = 2. At significance level .01, reject H_0 if $X^2 > 9.21$.

254

Computations:

State	Rearrested	Not rearrested	Total
California	217 (178.92)	69 (107.08)	286
Texas	278 (289.66)	185 (173.34)	463
Michigan	145 (171.42)	129 (102.58)	274
Total	640	383	1023

X^2 = 8.102 + 13.539 + 0.469 + 0.784 + 4.071 + 6.803
= 33.769

Since 33.769 > 9.21, the null hypothesis is rejected. The data contains sufficient evidence to conclude that the true proportion rearrested within three years of release is not the same for all three states.

16.25 H_0: There is no difference in mortality rate for the four concentrations.

       H_a: There is a difference in mortality rate for the four concentrations.

Test statistic: $X^2 = \sum \dfrac{(\text{observed count} - \text{expected count})^2}{\text{expected count}}$

Rejection region: For d.f. = (2 - 1)(4 - 1) = 3, Appendix Table XII gives the .01 level critical value as 11.34. The null hypothesis will be rejected if $X^2 > 11.34$.

Computations:

	Concentration				
	I	II	III	IV	Total
Survived	80 (74.5)	74 (74.5)	78 (74.5)	66 (74.5)	298
Died	0 (5.5)	6 (5.5)	2 (5.5)	14 (5.5)	22
Total	80	80	80	80	320

$X^2 = \dfrac{(80 - 74.5)^2}{74.5} + \dfrac{(74 - 74.5)^2}{74.5} + \ldots + \dfrac{(2 - 5.5)^2}{5.5} + \dfrac{(14 - 5.5)^2}{5.5}$
= .41 + .00 + .16 + .97 + 5.50 + .05 + 2.23 + 13.14 = 22.45

Since 22.45 > 11.34, the null hypothesis is rejected. The data strongly suggests that there is a difference in the mortality rate for the four concentrations.

16.27 H_0: Age at death and location of death are independent.

H_a: Age at death and location of death are dependent.

Test statistic: $X^2 = \sum \dfrac{(\text{observed count} - \text{expected count})^2}{\text{expected count}}$

Rejection region: For d.f. = (4 - 1)(3 - 1) = 6, Appendix Table XII gives the .01 level critical value as 16.81. The null hypothesis will be rejected if $X^2 > 16.81$.

Computations:

| | Location | | | |
Age	Home	Acute care	Chronic care	Total
15 - 54	94 (90.2)	418 (372.5)	23 (72.3)	535
55 - 64	116 (113.6)	524 (469.3)	34 (91.1)	674
65 - 74	156 (142.7)	581 (589)	109 (114.3)	846
Over 74	138 (157.5)	558 (650.3)	238 (126.2)	934
Total	504	2081	404	2989

$$X^2 = \frac{(94 - 90.2)^2}{90.2} + \frac{(418 - 372.5)^2}{372.5} + \ldots + \frac{(238 - 126.2)^2}{126.2}$$

$$= .16 + 5.56 + 33.63 + .05 + 6.39 + 35.79 + 1.25 + .11 + .25 + 2.41 + 13.09 + 98.94 = 197.62$$

Since 197.62 > 16.81, the null hypothesis is rejected. The data strongly suggests that the variables, age at death and location of death, are dependent.

16.29 a)

Cholesterol	Frequency	Expected count
less than 115	4] 10	4.58] 12.01
115 - < 135	6]	7.43]
135 - < 155	16	11.32
155 - < 175	11	11.74
175 - < 195	6	8.29
195 - < 215	2] 6	3.99] 5.64
215 or more	4]	1.65]
	n = 49	49

b) H_0: The population distribution of total serum cholesterol is normal.

H_a: The population distribution of total serum cholesterol is not normal.

Test statistic:
$$X^2 = \sum \frac{(\text{observed count - expected count})^2}{\text{expected count}}$$

Rejection region: For d.f. = 5 - 3 = 2, Appendix Table XII gives the .05 level critical value as 5.99. The null hypothesis will be rejected if $X^2 > 5.99$.

$$X^2 = \frac{(10 - 12.01)^2}{12.01} + \frac{(16 - 11.32)^2}{11.32} + \frac{(11 - 11.74)^2}{11.74}$$
$$+ \frac{(6 - 8.29)^2}{8.29} + \frac{(6 - 5.64)^2}{5.64}$$

$$= .34 + 1.93 + .05 + .63 + .02 = 2.97$$

Since 2.97 < 5.99, the null hypothesis is not rejected. The data suggests that the population distribution of total serum cholesterol is normal.

16.31 H_0: x has a binomial distribution with n = 7 and π = .5.
H_a: H_0 is not true.

Test statistic: $X^2 = \sum \frac{(\text{observed count - expected count})^2}{\text{expected count}}$

Rejection region: For d.f. = k - 1 = 7, Appendix Table XII gives the .01 level critical value as 18.48. The null hypothesis will be rejected if $X^2 > 18.48$.

Computations:

$$X^2 = \frac{(6 - 10.4)^2}{10.4} + \frac{(57 - 73)^2}{73} + \ldots + \frac{(13 - 10.4)^2}{10.04}$$
$$= 1.86 + 3.51 + .75 + .02 + .00 + 6.32 + .22 + .65$$
$$= 13.33$$

Since 13.33 < 18.48, the null hypothesis is not rejected. The data suggests that the variable, number of males in first seven children, has a binomial distribution with n = 7 and π = .5.

16.33 H_0: Number of Larrea divaricata plants follows a Poisson
        distribution.

   H_a: H_0 is not true.

   Test statistic: $X^2 = \sum \dfrac{(\text{observed count - expected count})^2}{\text{expected count}}$

   Rejection region: For d.f. = 5 - 2 = 3, Appendix Table XII gives
   the .05 level critical value as 7.82. The null hypothesis will
   be rejected if $X^2 > 7.82$.

   $X^2 = \dfrac{(9 - 5.9)^2}{5.9} + \dfrac{(9 - 12.3)^2}{12.3} + \dfrac{(10 - 13)^2}{13} + \dfrac{(14 - 9)^2}{9}$

   $\dfrac{(6 - 7.8)^2}{7.8} = 1.63 + .89 + .69 + 2.78 + .42 = 6.41$

   Since 6.41 < 7.82, the null hypothesis is not rejected. The data
   suggests that the variable number of Larrea divaricata plants
   follows a Poisson distribution.

Supplementary Exercises

16.35 H_0: Sex and seat belt usage are independent.
         H_a: Sex and seat belt usage are dependent.

Test statistic: $X^2 = \sum \dfrac{(\text{observed count} - \text{expected count})^2}{\text{expected count}}$

Rejection region: For d.f. = $(2 - 1)(2 - 1) = 1$ and Appendix Table XII gives the .05 level critical value as 3.84. The null hypothesis will be rejected if $X^2 > 3.84$.

Computations:

	Don't use	Use	Total
Male	192 (215.69)	272 (248.31)	464
Female	284 (260.31)	276 (299.69)	560
Total	476	548	1024

$$X^2 = \frac{(192 - 215.69)^2}{215.69} + \frac{(272 - 248.31)^2}{248.31} + \frac{(284 - 260.31)^2}{260.31}$$
$$+ \frac{(276 - 299.69)^2}{299.69}$$
$$= 2.60 + 2.26 + 2.16 + 1.87 = 8.89$$

Since $8.89 > 3.84$, the null hypothesis is rejected. The data suggests that the variables sex and seat belt usage are dependent.

16.37 H_0: Sex and relative importance assigned to work and home are independent.

         H_a: Sex and relative importance assigned to work and home are dependent.

Test statistic: $X^2 = \sum \dfrac{(\text{observed count} - \text{expected count})^2}{\text{expected count}}$

Rejection region: For d.f. = $(2 - 1)(3 - 1) = 2$, Appendix Table XII gives the .05 level critical value as 5.99. The null hypothesis will be rejected if $X^2 > 5.99$.

Computations:

Relative Importance

	work > home	work = home	work < home	Total
Female	68 (79.3)	26 (25.0)	94 (83.7)	188
Male	75 (63.7)	19 (20.0)	57 (67.3)	151
Total	143	45	151	339

$$X^2 = \frac{(68 - 79.3)^2}{79.3} + \frac{(26 - 25)^2}{25} + \ldots + \frac{(57 - 67.3)^2}{67.3}$$
$$= 1.61 + .04 + 1.26 + 2.01 + .05 + 1.56 = 6.54$$

Since 6.54 > 5.99, the null hypothesis is rejected. The data suggests that sex and relative importance assigned to work and home are dependent variables.

16.39 H_0: The variable size of farm and oldest child's residence are independent.

H_a: The variable size of farm and oldest child's residence are dependent.

Test statistic: $X^2 = \sum \frac{(\text{observed count} - \text{expected count})^2}{\text{expected count}}$

Rejection region: For d.f. = (3 - 1)(2 - 1) = 2, Appendix Table XII gives the .05 level critical value as 5.99. The null hypothesis will be rejected if $X^2 > 5.99$.

Computations:

	With parents	Separate residence	Total
0 -< .3	28 (21.9)	8 (14.1)	36
.3 -< 1.0	10 (11.0)	8 (7.0)	18
1.0 -< 9.0	18 (23.1)	20 (14.9)	38
Total	56	36	92

$$X^2 = \frac{(28 - 21.9)^2}{21.9} + \frac{(8 - 14.1)^2}{14.1} + \ldots + \frac{(20 - 14.9)^2}{14.9}$$
$$= 1.69 + 2.63 + .08 + .13 + 1.14 + 1.77 = 7.44$$

Since 7.44 > 5.99, the null hypothesis is rejected. The data suggests that size of farm and oldest child's residence are dependent variables.

16.41 H_0: Age and rate believed attainable are independent variables.

H_a: Age and rate believed attainable are dependent variables.

Test statistic: $X^2 = \sum \dfrac{(\text{observed count - expected count})^2}{\text{expected count}}$

Rejection region: For d.f. = (4 - 1)(4 - 1) = 9, Appendix Table XII gives the .01 level critical value as 21.67. The null hypothesis will be rejected if $X^2 > 21.67$.

Computations:

Rates believed attainable

		0 - 5	6 - 10	11 - 15	Over 15	Total
	Under 45	15 (28.4)	51 (72.1)	51 (28.2)	29 (17.3)	146
	45 - 54	31 (54.8)	133 (139.3)	70 (54.5)	48 (33.4)	282
Age	55 - 64	59 (49.2)	139 (124.9)	35 (48.9)	20 (29.9)	253
	65 over	84 (56.6)	157 (143.7)	32 (56.3)	18 (34.4)	291
	Total	189	480	188	115	972

$$X^2 = \frac{(15 - 28.4)^2}{28.4} + \frac{(51 - 72.1)^2}{72.1} + \ldots + \frac{(18 - 34.4)^2}{34.4}$$

= 6.31 + 6.17 + 18.35 + 7.96 + 10.36 + .28 + 4.38 + 6.42
+ 1.95 + 1.58 + 3.97 + 3.3 + 13.28 + 1.23 + 10.48 + 7.84
= 103.87

Since 103.87 > 21.67, the null hypothesis is rejected. The data very strongly suggests that the variables, age and rate believed attainable are dependent variables.

16.43 H_0: The true proportions of individuals in each of the cocaine use categories do not differ for the three treatments.

H_a: The true proportions of individuals in each of the cocaine use categories differ for the three treatments.

Test statistic: $X^2 = \sum \dfrac{(\text{observed count - expected count})^2}{\text{expected count}}$

Rejection region: For d.f. = (4 - 1)(3 - 1) = 6 and Appendix Table XII gives the .05 level critical value as 12.59. The null hypothesis will be rejected if $X^2 > 12.59$.

Supplementary Exercises

Computations:

Treatment

Usage	A	B	C	Total
None	149 (118.9)	75 (84.8)	8 (28.3)	232
1 - 2	26 (34.8)	27 (24.9)	15 (8.3)	68
3 - 6	6 (19.0)	20 (13.5)	11 (4.5)	37
7 or more	4 (12.3)	10 (8.8)	10 (2.9)	24
Total	185	132	44	361

$$X^2 = \frac{(149 - 118.9)^2}{118.9} + \frac{(75 - 84.8)^2}{84.8} + \ldots + \frac{(10 - 2.9)^2}{2.9}$$

$$= 7.62 + 1.14 + 14.54 + 2.25 + .18 + 5.44 + 8.86 + 3.1 + 9.34$$
$$+ 5.6 + .17 + 17.11 = 75.35$$

Note: There are two cells with expected counts of 5 or less. If you combine the two usage categories of 3 - 6 and 7 or more the following analysis results.

Treatment

Usage	A	B	C	Total
none	149 (118.9)	75 (84.8)	8 (28.3)	232
1 - 2	26 (34.8)	27 (24.9)	15 (8.3)	68
3 or more	10 (31.3)	30 (22.3)	21 (7.4)	61
Total	185	132	44	361

$$X^2 = \frac{(149 - 118.9)^2}{118.9} + \frac{(75 - 84.8)^2}{84.8} + \ldots + \frac{(21 - 7.4)^2}{7.4}$$

$$= 7.62 + 1.14 + 14.54 + 2.25 + .18 + 5.44 + 14.46 + 2.65$$
$$+ 24.75 = 73.03$$

The d.f. = $(3 - 1)(3 - 1) = 4$ and the .05 level critical value is 9.49.

In either analysis, the null hypothesis is rejected. The data suggests the true proportions of individuals in each of the cocaine use categories differ for the three treatments.

16.45 H_0: Age and need for item pricing are independent variables.
H_a: Age and need for item pricing are dependent variables.

Test statistic: $X^2 = \sum \frac{(\text{observed count - expected count})^2}{\text{expected count}}$

Rejection region: For d.f. = (2 - 1)(5- 1) = 4 and Appendix Table XII gives the .05 level critical value as 9.49. The null hypothesis will be rejected if $X^2 > 9.49$.

Computations:

	Age					
	< 30	30 - 39	40 - 49	50 - 59	≥ 60	Total
Want	127 (131.1)	118 (123.3)	77 (71.7)	61 (55.1)	41 (42.8)	424
Don't Want	23 (18.9)	23 (17.7)	5 (10.3)	2 (7.9)	8 (6.2)	61
Total	150	141	82	63	49	485

$$X^2 = \frac{(127 - 131.1)^2}{131.1} + \frac{(118 - 123.3)^2}{123.3} + \ldots + \frac{(8 - 6.2)^2}{6.2}$$
$$= .13 + .22 + .39 + .64 + .08 + .91 + 1.56 + 2.74 + 4.43 + .55 = 11.65$$

Since 11.65 > 9.49, the null hypothesis is rejected. The data suggests that the variables age and need for item pressing are dependent.